DUNK YOUR BISCUIT HORIZONTALLY

106 Strange Scientific Facts

Biscuit

Dry
Sudden

Tea

Mug

fig. 1: The correct angle for
dunking a biscuit

Rik Kuiper and Tonie Mudde

summersdale

DUNK YOUR BISCUIT HORIZONTALLY

This edition published in 2010 by Summersdale Publishers Ltd.

First published by Uitgeverij Podium, NL in 2007 as KIJK NIET NAAR MOOIE VROUWEN (ALS JE NOG MOET NADENKEN)

Summersdale Publishers Ltd
46 West Street
Chichester
West Sussex
PO19 1RP
UK

www.summersdale.com

Printed and bound in Great Britain

ISBN: 978-1-84953-053-8

Substantial discounts on bulk quantities of Summersdale books are available to corporations, professional associations and other organisations. For details contact Summersdale Publishers by telephone: +44 (0) 1243 771107, fax: +44 (0) 1243 786300 or email: nicky@summersdale.com.

CONTENTS

Suck Up to Your Boss
Ten Tips for the Office..5

While Cycling, Wear a Wig, Not a Helmet
Nine Tips for on the Road...21

Drink Beer (and Save the Environment)
Six Tips for a Night Out..37

Talk About Hooligans (and Win Trivial Pursuit)
Seven Tips for your Spare Time..............................47

First Kiss? Tilt Your Head to the Right
Fourteen Tips for a Better Love Life........................59

Brush Your Teeth (and Reduce the Risk of a Heart Attack)
Six Tips for a Healthy Body......................................79

Spread Lemon Scent (and Everybody Will Start Cleaning)
Four Tips for in and Around the House.................89

Dunk Your Biscuit Horizontally, Not Vertically
Twelve Tips for Eating and Drinking......................................97

Think About Weight Lifting (and Get in Shape)
Six Tips for Sports Fans...115

Don't Look at Beautiful Women (If You Still Have Thinking to Do)
Five Tips for Small and Large Purchases.............................125

Ask For a Rectal Massage (and Get Rid of Those Hiccups)
Thirteen Tips to Overcome Pain and Other Discomforts.....135

Wash Your Hands (Cleanse Your Soul)
Five Tips to Get Rid of Your Traumas and Other Psychological Problems...155

Vicious Dog? Don't Trust the Owner
Nine Tips That Are Certainly Useful, But Did Not Fit into the Other Chapters..165

SUCK UP TO YOUR BOSS

TEN TIPS FOR THE OFFICE

DON'T SEND GROUP EMAILS

Got an urgent question? Send it to ten people simultaneously! That way you increase your chances of a speedy reply. But is that really true? Researchers from the Israel Institute of Technology in Haifa invented a nice girl: Sarah Feldman. They set up a free email address for her on Yahoo! and let her send a message with the subject matter 'Please help' to various university employees. Her question was simple: has the university also got a biology department? Every recipient should have been able to answer this question: yes there is. Some emails Sarah sent to five recipients simultaneously, others she sent to an individual.

The result? Of the recipients of the group email, just over half answered. But of those who had been addressed personally nearly two-thirds answered. What's more, they used significantly more words in their replies and more often gave extra information, like telephone numbers or a web address. In short: the more addressees an email has the less likely a satisfactory response. And why is that? Because people feel less responsible if there are other recipients. Luckily the solution is simple. Still want to address several people simultaneously with the same request? Then do it in separate emails.

Barron, G. & Yechiam, E. 'Private e-mail requests and the diffusion of responsibility' in *Computers in Human Behaviour* 18, 5 (2002), pp. 507–520

SUCK UP TO YOUR BOSS

Sucking up works, an experiment conducted by Roos Vonk, Professor of Psychology at Radboud University in Nijmegen, shows. In her research volunteers were given the role of manager and asked to complete a detailed personality test. In it they were asked how they felt about values such as friendship and wealth. Later they received reactions to their completed tests from their subordinates. At least, that is how it was presented to them. In fact there were no subordinates and all the volunteers received the same letter. That letter was full of flattering comments like:

'*I think everyone would get along well with him, and I certainly would.*'
'*I think she would deliver great work in various areas, because she has so many talents.*'

The managers were then asked to rate the subordinates anonymously. How sincere did they find the letter-writer? How friendly? How slimy? And now for the crux. While a section of the managers thought they were reading a reaction to their own personality tests, another section was told that the letter was a reaction to other managers' tests. What was revealed? Volunteers who thought they were reading about themselves found the subordinate kinder, more sincere and less slimy that those who thought they were reading about others. Conclusion: as long as the sucking up is aimed at us, we think it's great. But if the exact same flattering comments are aimed at another, we think: what a slime ball!

Vonk, R. 'The slime effect: suspicion and dislike of likeable behaviors towards superiors' in *Journal of Personality and Social Psychology* 74 (1998), pp. 849–864

PUT EYES ON THE WALL (AND EARN MORE)

Let's say that next to the departmental coffee machine at work there stands a jar or a piggy bank. And stuck on the wall is a list of prices with the request to pay for the coffee consumed by putting the correct amount into the jar/piggy bank. But every week the takings fall short. It seems many are inclined to walk away without paying. After all, no one controls it.

British research reveals that there is a way to make sure that such shortfalls in the workplace are reduced. Psychologist Melissa Bateson at the University of Newcastle put a new drinks price list up every week in her own department. She kept the price of tea, coffee and milk the same but changed the picture above the list. One week she chose a picture of flowers, the next a picture of eyes that looked straight at you. In the weeks that the coffee drinkers were looked at by a pair of photocopied eyes, the contributions were at least two and a half times greater than when flowers were on the wall. The reason is simple according to Bateson: if people think they are being watched, they are more honest.

Bateson, M., Nettle, D. & Roberts, G. 'Cues of being watched enhance cooperation in a real-world setting' in *Biology Letters* 2 (2006), pp. 412–414

4.

DON'T USE INTERESTING FONTS

Finally, you've finished that report. All that remains is to choose an interesting font. But be careful; no matter how well your account is put together, this is the moment you can come a cropper. At least, so say the results of an experiment by Daniel Oppenheimer of Princeton University in the USA. The psychologist gave 51 students an identical text, written in one of the following fonts:

The world famous Times New Roman.

And the less common Monotype Corsiva.

The students had to give marks out of seven for how intelligent they thought the author was. And lo and behold, the author who had chosen the Monotype Corsiva got an average of half a point lower than the author who had written the same report in Times New Roman.

A similar experiment by Oppenheimer provides another useful tip: don't use complicated synonyms to seem more intelligent. For example, avoid 'comprehend' and simply use 'understand'. Readers rate the author of complex prose as less intelligent than the author of the same text written in simple terms.

Oppenheimer, D. M. 'Consequences of erudite vernacular utilized irrespective of necessity: problems with using long words needlessly' in *Applied Cognitive Psychology* 20 (2006), pp. 139–156

5.

STICK A POST-IT ON THE SURVEY

When someone sends out a survey, they obviously want as many people as possible to respond. Attaching a Post-it to the survey does wonders, experiments by Psychologist Randy Garner of Sam Houston State University in the USA show. He sent 150 university colleagues a five-page satisfaction survey, with questions like: 'What do you think of the facilities on campus?', 'How high do you put the reputation of the university?', etc.

The same question appeared on all the surveys: 'Do you have five minutes to complete this survey? Thank you!' On 50 surveys this question was also written by hand on the front page. And on 50 others on a Post-it, stuck on the front page. After giving recipients two weeks to send in the completed surveys, Garner noted the results.

Request Type	Percentage returned
No handwritten request	36 %
Handwritten request on front page	48 %
Handwritten request on a Post-it	76 %

Now Garner was faced with the question: is the high response because of the yellow sticker itself, or because of the combination of text and Post-it?

He investigated in a further experiment, sending surveys out with blank Post-its. The response was clearly lower. So researchers take note: if you want a better response rate to your survey, make the effort to write a few friendly words on that sticky piece of paper.

Garner, R. 'Post-it note persuasion: a sticky influence' in *Journal of Consumer Psychology* 15 (2006), pp. 230–237

6.

STAY SILENT DURING BRAINSTORMING SESSIONS

'Time for a brainstorm!' your boss calls, and before you know it you are sitting at a table with the whole department throwing ideas around. Stop this, says the social psychologist, Bernard Nijstad. For his PhD at Utrecht University, Netherlands, he organised different kinds of brainstorming sessions, with independent juries who gauged the number of good ideas produced. The result? In the traditional brainstorm, where everyone can shout out ideas, scores were markedly low.

The traditional system requires that participants constantly check who is speaking, so they can't concentrate fully on generating ideas. And if someone has a good idea, they have to wait their turn to speak, with the risk that the idea fades before they have chance to articulate it.

Conclusion? The best brainstorm is one where everyone keeps their mouths shut. But then how can you react to others' ideas? Nijstad discovered that the following method, which he calls 'brain-write', works very well:

- First think of a well-phrased question, for example: how can we make our product more attractive to a younger market?
- Then put a pile of blank pages in the middle of the table.
- Everyone takes a page and writes an idea down. Ready? Slide it back to the middle, pick up another page at random and write another idea down, letting yourself be inspired by the other person's idea on the page. Make sure that there a few more pages than participants, to ensure there is always an available page.
- Stop after 10 minutes: enough time to generate a couple of brilliant ideas.

An added bonus of this technique is anonymity. Participants don't know if the idea is the boss's or a work experience person's, and so take all the ideas equally seriously.

Nijstad, B. A. 'How the Group Affects the Mind: Effects of Communication in Idea-Generating Groups' (2000), PhD Utrecht University

PAY ATTENTION TO SOMEONE'S HANDSHAKE (AND FEEL THEIR INSECURITY)

What we all already suspected has now finally been scientifically proven: he who has a weak handshake, has a weak character. At the University of Alabama, four people underwent a month's training on how to judge handshakes. A tough task, because during the short period of a handshake the four had to note strength, humidity, length, eye contact and 'the overall grip'. After the training they shook hands with 112 unsuspecting students. These test cases had earlier that day filled in a personality test. Result? Shy students gave a weak handshake while extrovert students gave a firm handshake. This was true for men and women.

Brown, J. D., Chaplin, W. F., Clanton, N. R., Philips, J. B. & Stein, J. L. 'Handshaking, gender, personality and first impressions' in *Journal of Personality and Social Psychology* 79 (2000), pp. 110–117

SHOW THEM YOUR RIGHT CHEEK (AND LOOK INTELLIGENT)

Holland's Leiden University has a large collection of portraits of professors. Most are so-called 'three-quarter' portraits, with one cheek more prominent than the other. In Leiden the Dutch behavioural biologist Carel ten Cate examined four centuries of posing professors and discovered a remarkable pattern: most of the scientists had turned their right cheek to the audience. Why? In his search to find out, Cate came across an interesting Australian experiment. Before being photographed, volunteers were either told that they were posing for their family, or for the Australian National Academy of Science. When posing for their relatives the participants unconsciously preferred to turn their left cheek to the camera. Those who posed for the scientific community showed more right cheek. The explanation probably lies in our brain. Most of our emotions are processed in the right side of the brain. Because this side controls the left side of the face, showing some left cheek is best for expressing emotions. The left side of the brain, which controls the right side of the face, is better at analytical processes. So by slightly turning our head we either emphasise our emotional or our analytical side.

One question remains unanswered: do we really look more intelligent when we show people our right cheek. Cate asked more than 80 volunteers to give grades to a large number of portrait photos. On a scale of one to six the participants had to judge how 'scientific' they thought the people in the photos looked. Final result? Yes, we look more intelligent when we show people our right cheek.

ten Cate, C. 'Posing as professor: laterality in posing orientation for portraits of scientists' in *Journal of Nonverbal Behavior* 26, 3 (2002), pp. 175–192

9.

LEAN BACK IN YOUR CHAIR

Sitting up nice and straight in your chair is best for your back, isn't it? Wrong, says research from the University of Alberta in Canada. Healthy participants without back complaints were given an MRI scan, an apparatus that assesses the body with magnetic rays. Normally these scanners are tunnel-like and close in tightly around the prostrate person. For this experiment a special model big enough to sit in was used. The volunteers sat in a chair in three different positions: slightly bent forward, perfectly straight and leaning casually back. The researchers took note when analysing the MRI scans of, amongst other things, the thickness and movement of intervertebral discs. Conclusion: the stress on the spine is greatest when the back and upper legs make a 90 degree angle, sitting straight up in other words. Leaning forward is bad too, because the lower back is put under great pressure. Leaning back slightly so that the back and upper legs make an angle of 135 degrees is the best posture.

Bashir, W. & Torio, T. 'The way you sit will never be the same! Alterations of lumbosacral curvature and intervertebral disc morphology in normal subjects in variable sitting positions using whole-body positional MRI', Annual Meeting of Radiological Society of North America 2006

CHEW GUM (AND REMEMBER MORE)

It must have been a strange sight to behold; dozens of jaws grinding during the chewing gum experiment conducted by Northumbria University in the UK. Seventy-five healthy volunteers were split into three groups:

- Group 1 – chewed a sugar-free gum: Wrigley's Extra Spearmint.
- Group 2 – imitated chewing actions but had empty mouths.
- Group 3 – had no chewing gum and kept their jaws still.

During the experiment the volunteers did a series of memory tests. What happened? Those in Group 1, who were chewing gum, remembered the most words, both in the short term and the long term. This came to 35 per cent more than the volunteers in Group 3, who kept their jaws locked. Is there a logical explanation? Yes, although it hasn't been 100 per cent proven. Chewing increases the heart rate. This leads to more oxygen and nutrition being pumped to the brain, enabling it to process information more efficiently. On top of that the body creates insulin while chewing. This hormone stimulates the hippocampus, a part of the brain that plays an important role in memory.

Strangely, the 'fake-chewers' of Group 2 performed the least well of the three groups in the memory test. The researchers suspect this is due to the unnatural activity of chewing when they had nothing in their mouths: they were probably too busy reminding themselves to keep chewing.

Scholey, A., Wesnes, K. & Wilkinson, L. 'Chewing gum selectively improves aspects of memory in healthy volunteers' in *Appetite* 38 (2002), pp. 235–236

WHILE CYCLING, WEAR A WIG, NOT A HELMET

NINE TIPS FOR ON THE ROAD

DON'T WASTE MONEY ON AN AIRBAG

Buying a car? If you care about your safety you will of course invest in an airbag. And hey, you might as well add an anti-lock braking system (ABS), to prevent the wheels slipping during an emergency stop. According to laboratory evidence these things increase safety dramatically: due to ABS an accident is less likely, and should something happen the airbag prevents injury.

But is it really worth the investment? This was the question asked by Professor of Civil Engineering Fred Mannering of Purdue University in the USA. Together with other researchers he analysed the characteristics of drivers driving around the state of Washington between 1992 and 1996. Did the owners of cars with airbags or anti-lock brakes have less accidents? No. And what's more, their physical injuries turned out to be just as bad as for those with less 'safe' cars. For the researchers this result did not come as a big surprise. It is often the case that extra safety measures lead people to take bigger risks. This is known as the 'offset hypothesis': drivers with airbags or ABS feel safer and so keep less distance, change lanes more often on the motorway and take more risks at junctions than when they don't have the extra features.

Maheshri, V., Mannering, F. L. & Winston, C. 'An exploration of the offset hypothesis using disaggregate data: the case of airbags and antilock brakes' in *Journal of Risk and Uncertainty* 32 (2006), pp. 83–99

NO SEAT? JUST ASK

Failed to get a seat on the train yet again? Don't stand there sulking, just ask someone: 'Excuse me, do you think I could sit down?' It is very likely that the person will relinquish his/her seat. It was Stanley Milgram who discovered this phenomenon in 1971. One day his mother-in-law was complaining that young people no longer get up for the old on buses. 'Did you ever ask one of them for a seat?' asked the famous psychologist. 'No' she said. So Milgram decided to ask his students to take part in an experiment. In the New York metro they would ask sitting passengers to give up their place. The students laughed heartedly at the idea. Only one volunteered. But the results this young man delivered were staggering. He was regularly offered a seat.

Milgram decided to broaden the experiment: he sent more students out. The results were surprising. At least 56 per cent of the passengers gave up their seat without the students having to give a good reason why they wanted to sit. And a further 12 per cent shifted along so that the student could fit in beside them. The reason, Milgram suspected, is that people feel threatened. Giving up a seat is easier than passing over the request. The biggest problem with this technique, as the psychologist discovered when he stepped into the metro himself, is the indescribable angst when asking this simple question.

Milgram, S. & Sabini, J. 'On maintaining urban norms: a field experiment in the subway' in Baum, A., Singer, J. E. & Valins, S. *Advances in Environmental Psychology* 1 (1978), pp. 31–40

Blass, T. *The Man Who Shocked the World: The Life and Legacy of Stanley Milgram* (2004, Basic Books).

DON'T WEAR A HOODY IN TRAFFIC

Anyone who has seen the American cartoon series *South Park* knows that the character Kenny dies in every episode. His most important identifying feature: he wears an orange parka with an enormous hood. Only his big round eyes are visible. British research shows that the connections between accidents and hoodies isn't that far fetched. A large hoody, fastened tightly, causes a narrowing of the field of vision and that can cause dangerous situations in traffic. To assess the level of danger scientists invited six healthy adults for a check up at the Birmingham and Midland Eye Centre. They first established the average field of vision of the volunteer without hoody. Then they asked them to put on four different anoraks one after another. The strings of the hoody were pulled tight. That created acute tunnel vision: from left to right the field of vision was reduced by a third, up and down by nearly two thirds. According to researchers it would be even worse in practice, because when it is raining people also keep their heads down. So how should a cyclist keep

25

dry? Unfortunately, the researchers didn't provide an answer to that.

Cheung, C. M. G. & Durrani, O. M. 'The danger of wearing an anorak' in *Journal of the Royal Society of Medicine* 95 (2002), pp. 192–193

14.

SPARE YOUR LUNGS – TAKE THE CAR

We can't see it but it floats all around us: fine dust. The particles are less than 100 nanometres in diameter, are 800 times smaller than the thickness of a hair and can cause heart and lung diseases. So how can we make sure we inhale as little fine dust as possible? Luckily, research from Imperial College London offers a helping hand. Researchers took five different modes of transport in busy London and measured the quantity of fine dust to which they were exposed on each type. Here are the results:

Mode of Transport	Number of particles per cm³ of air
Car	40,000
Walking	50,000
Bicycle	80,000
Bus	90,000
Taxi	100,000

So why did the car have the best score? Thanks to the closed windows and ventilation system, car drivers and passengers don't come into direct contact with the outside air. This advantage doesn't hold for taxis or buses: due to the enormous quantity of driving hours in the busy city fine dust gets piled up inside them. And why do cyclists take in more dust than walkers? Because cycle wheels are that much closer to exhaust fumes; those couple of metres make a real difference. Pedestrians who walk alongside buildings breathe in 10 per cent less fine dust than those who walk close to the curb.

Arnold, S. J., Clark, R. D. R., Colvile, R. N., Kaur, S., Nieuwenhuijsen, M. & Walsh, P. T. 'Exposure visualisation of ultrafine particle counts in a transport microenvironment' in *Atmospheric Environment* 40, 2 (2006), pp. 386–398

15.

MEN ONLY
USE A WOMAN'S SADDLE (OR A RECUMBENT CYCLE)

Cyclists don't talk about it much, but many of them suffer erectile problems. A survey of 1,786 amateurs in and near the German town of Cologne showed that this problem occurs three times more often in male cyclists than in men who don't cycle. Why is this? Frank Sommer of the local teaching hospital set up an experiment in order to find out. He invited sporty men to the lab and asked them to mount an exercise bike for half an hour's training. In order to measure the oxygen flowing to the genitals, he placed an electrode on the head of each man's penis. During cycling the blood flow to the penis was reduced significantly. Some of the men were asked to continue cycling in a standing position after 15 minutes; in these instances the blood flow to the genitals rose immediately.

So what was going on? When sitting on a saddle, all of a cyclist's weight rests on the perineum, the area between the anus and the testicles. This is precisely the place where the blood vessels and nerves that feed the penis are located, so the genitals get less oxygen. For men who cycle often this can lead to poorly functioning swelling bodies: erection problems. Sommer recommends that cycling fanatics use a broader saddle, so the weight is spread more evenly. From comparative saddle research it seems a woman's saddle (an oval-shaped saddle with no nose) causes the least damage. Cyclists should also stand on the pedals every 10 minutes or so. But the ideal solution would be to swap the racing bike for a recumbent bike – that way the penis gets pretty much the same oxygen as usual.

Sommer, F. 'Die Einflüsse des Fahrradfahrens auf die männliche Sexualität, Teil 1: erektile Dysfunktion und Fahrradfahren' in *Blickpunkt der Mann* 2, 1 (2004), pp. 28–32

Salomon, G. & Sommer, F. 'Der Einfluss des Fahrradfahrens auf die männliche Sexualität: auswirkungen auf die Erekktionsfunction', in *Journal für Reproduktionsmedizin und Endokrinologie* 3, 3 (2006), pp. 141–144

16.

STARE AT THE DRIVER (AND GET A LIFT SOONER)

You're standing at the roadside, thumb out. Is there any way of increasing your chances of getting a lift? Various American scientists asked themselves this very question back in the 1970s. Here are some of their conclusions:

- Psychologist Mark Snyder of Minnesota University analysed who had more success: a man, a woman or a man and woman together. The woman alone was the most successful. A couple got a lift as often as a man alone.

- Then Snyder told the hitchhikers to look the person behind the wheel in the eye for as long as possible. That seemed to help. Those who looked drivers in the eye got more than twice as many lifts as the hitchers who looked up at the sky or down at their feet.

- What should a hitcher wear? Peter Crassweller of the Southern Methodist University in Dallas dressed his hitcher first as a hippy: long hair and a bandana, flairs, no shoes or socks, a fag hanging from the corner of his mouth. Then the same young man put on a suit and tie. In pretty much every location the smartly dressed version was picked up sooner than the hippy.

- Charles Morgan of Washington University, Seattle, found that women who accentuated their cleavage and looked the driver in the eye were picked up much sooner. Roughly one in ten drivers stopped. A woman who made no eye contact and whose breasts were less obvious had to wait twice as long.

Crassweller, P., Gordon, M. A. & Tedford Jr, W. H. 'An experimental investigation of hitchhiking' in *The Journal of Psychology* 82 (1972), pp. 43–47

Grether, J., Keller, K. & Snyder, M. 'Staring and compliance: a field experiment on hitchhiking' in *Journal of Applied Psychology* 4 (1974), pp. 165–170

Fahrenbruch, C. E., Locard, J. S., Morgan, C. J. & Smith, J. L. 'Hitchhiking: social signs at a distance' in *Bulletin of the Pschynomic Society* 5 (1975), pp. 459–461

17.

WHILE CYCLING, WEAR A WIG, NOT A HELMET

There he goes, Ian Walker from the University of Bath, on his bicycle. The traffic psychologist attached a video camera to his handlebars, and to his saddle a device to measure the distance between him and the overtaking traffic. Dressed in normal clothes, he cycled a total of 320 kilometres through the cities of Bristol and Salisbury. Sometimes he would wear a white helmet, at other times he went bare-headed. And on a couple of trips the bearded researcher wore a black wig, which made the traffic approaching him from behind think he was a woman.

Afterwards Walker analysed data from more than 3,000 cars that had passed him while cycling. To his astonishment, the drivers kept less distance when he was wearing a helmet. Walker presumes that when drivers can see that a cyclist has the extra protection of a helmet, it unconsciously tempts them to take bigger risks. Remarkably, when he cycled wearing the wig, the cars kept a bigger distance. Possibly drivers think that women are more vulnerable or unpredictable than men, the cycling researcher suggests.

Interesting detail: while on the road Walker had two collisions. Both times he wore a helmet.

Walker, I. 'Drivers overtaking bicyclists: objective data on the effects of riding position, helmet use, vehicle type and apparent gender' in *Accident Analysis and Prevention* 39 (2007), pp. 417–425

DON'T EVEN PHONE HANDS-FREE

If you are caught driving with a mobile pressed to your ear, you can count on a substantial fine. However, calling hands-free is allowed. This is completely unjust according to researchers at the University of Utah.

A study was set up in which experienced drivers were placed in a simulator and told to follow a virtual car. Whilst driving, they were also required to speak on a mobile phone, either with the phone to their ear or through a hands-free kit. Volunteers in a control group drove without distractions. The drivers' task was simple: as soon as the virtual car slowed down they had to slam on the brakes. What transpired? The drivers who were speaking on mobile phones, even if they were using a hands-free kit, braked 9 per cent later than those who weren't on the phone. During the experiment three of the 40 drivers in the mobile phone group even crashed into the car they were following.

In another group, drivers without a mobile phone were given vodka to drink before beginning the simulation. It turned out that the performance of those on the phone was even worse than of those drivers who had necked a couple of glasses of vodka before getting behind the wheel!

Crouch, D. J., Drews, F. A. & Strayer, D. L. 'A comparison of the cell phone driver and the drunk driver' in AEI-Brookings Joint Centre Working paper 04-13 (2004)

Boer, E., Levy, J. & Pashler, H. , 'Central interference in driving: is there any stopping the psychological refractory period?' in *Psychological Science* 17, 3 (2006), pp. 228–235

19.

DON'T LISTEN TO MUSIC WHILE FLYING

In many places music lovers readily put the volume of their MP3 players a few notches higher. But an experiment conducted by Harvard Medical school, amongst others, shows that this is a dangerous habit. The researchers recorded the noise in an airplane cabin and played it via a speaker system into a lab. In this lab volunteers listened with various types of headphones to MP3 players. They could control the volume themselves. During the 'flight' 80 per cent of the volunteers put the volume of their MP3 players dangerously high. This was true for the in-ear headphones as well as the on-ear headphone models. Listening to music at too high a volume can lead to permanent hearing damage, because the vibrations cause hair cells in the inner ear to break. In quiet environments the volunteers kept the volume of their MP3 players under control: only 6 per cent pounded too many decibels against their eardrums.

Those who still want to listen to music in noisy places would do well to buy special headphones that reduce the surrounding noise. With those a 'mere' 20 per cent wreck their ears.

% max volume	Max listening time to avoid hearing damage (with an MP3 player with standard earphones)
10–50%	no limit
60%	18 hours
70%	4 hours 36 mins
80%	1 hour 12 mins
90%	18 mins
100%	5 mins

Fligor, B. J. & Ives, T. 'Does earphone type affect risk for recreational noise-induced hearing loss?' National Hearing Conservation Association (2006)

DRINK BEER (AND SAVE THE ENVIRONMENT)

SIX TIPS FOR A NIGHT OUT

BE ON THE ALERT FOR DRINKING SONGS

Are there regular sing-alongs in your local? Then keep your eyes on your wallet, because clever pub landlords might well be in on the results of research by Celine Jacob of Rennes University, France.

Armed with a stopwatch and notepad the researcher sat unobtrusively in the corner of a small bar in a French seaside town for 14 days, from 2 p.m. to 4 p.m. every afternoon. She observed a total of 93 clients. For each she noted the hour of arrival and departure and the size of the bill. At her request the music was changed from time to time. Sometimes the speakers were pumping out top 40 hits, sometimes songs from cartoon films and sometimes drinking songs. If they heard drinking songs, the patrons stayed longer and spent more; on average nearly five euros in some 21 minutes. Why was this? Because the pub frequenters associated those songs with a good atmosphere and drinking, Jacob concluded.

Jacob, C. 'Styles of background music and consumption in a bar: an empirical evaluation' in *International Journal of Hospitality Management* 25, 4 (2006), pp. 716–720

21.

DRINK BEER (AND SAVE THE ENVIRONMENT)

Could drinking beer help keep our rivers clean? Apparently so, according to research by the Japanese Kobe Pharmaceutical University. The scientists discovered that a by-product of the beer-brewing process – the finely powdered skin of grain known as bran – can be used to clean polluted water.

The researchers demonstrated its purifying properties on dangerous and carcinogenic industrial waste. Currently many factories purify their waste water with active coal filters ('active coal' for those in the know). The same black granules are in anti-diarrhoea tablets. Thanks to their many small pores they block all sorts of pollutants. The downside is that active coal is very expensive and energy intensive to create. So from now on we'd be better off using bran, the Japanese researchers think. After all, we get it for free when brewing a keg of beer. And what's more it filters better than active coal.

It makes a good argument for staying longer down the pub: 'Lets have another one. For the environment.'

Adaschi, A., Kasuga, I., Okano, T. & Ozaki, H. 'Use of beer bran as an absorbent for the removal of organic compounds from wastewater' in *Journal of Agricultural and Food Chemistry* (2006)

DON'T TRUST ANY HANGOVER CURES

Is there anything that really helps get rid of a hangover? Max Pittler of the Peninsula Medical School, part of the Universities of Exeter and Plymouth found some extraordinary suggestions on the Internet. Aspirin is recommended, as are bananas, coal, charcoal tablets, eggs, green tea, a hot bath, a milkshake, a pizza or the good old-fashioned hair of the dog.

But are the claimed benefits of such remedies scientifically provable? Pittler and colleagues scrutinised the medical literature and found eight serious studies on anti-hangover methods. They looked into the medicines tropisetron and propanolol and household goods like cucumbers, artichokes and yeast solution. The surprise result? None of the miracle cures turned out to work. And so the researchers concluded that there is only one thing that actually helps: drinking less.

Ernst, E., Pittler, M. H. & Verster, J. C. 'Interventions for preventing or treating alcohol hangover: systematic review of randomised controlled trials' in *British Medical Journal* 331 (2005), pp. 1515–1518

SMOKE A COUPLE OF CIGARETTES (AND GET DRUNK LESS QUICKLY)

We all know that smoking is bad for our health. But it can help you get through a breathalyser test. The Texan neurologist Wei-Jung Chen administered doses of nicotine and alcohol to adult rats. He then took some blood from their tails and analysed the blood alcohol level. And what did he find? A high dose of nicotine slows down drunkenness.

Previously, the Royal Adelaide Hospital in Australia had already shown that this also applies to humans. Eight heavy smokers (20–35 cigarettes per day) smoked four cigarettes in an hour and then drank two glasses of a light alcoholic drink. They drank the same amount of alcohol after seven days of no smoking. In both cases the blood alcohol level was measured half an hour after consumption. The final score: after smoking four cigarettes the blood alcohol level was nearly 18 per cent lower than after the period of non-smoking. How is this is possible? Nicotine delays gastric emptying. Because of the delay, the alcohol has time to be broken down and so it does not get absorbed into the bloodstream of smokers.

Chen, W. J. A., Parnell, S. E. & West, J. R. 'Nicotine decreases blood alcohol concentrations in adult rats: a phenomenon potentially related to gastric function' in *Alcoholism: Clinical and Experimental Research* 30, 8 (2006), pp. 1408–1413

Johnson, R. D., Horowitz, M., Maddox, A. F., Shearman, D. J. C. & Wishart, J. M. 'Cigarette smoking and rate of gastric emptying: effect on alcohol absorption' in *British Medical Journal* 302 (1991), pp. 20–23

DON'T MIX VODKA AND RED BULL (IT WON'T GIVE YOU WINGS)

Mixing some Red Bull into your spirits keeps you going for longer on the dance floor, doesn't it? A team of researchers from the University of Sao Paulo in Brazil would beg to differ.

A group of healthy men was given a new drink every week. Sometimes it contained alcohol, sometimes alcohol and Red Bull. An hour after consuming the drink they got on bicycles and pedalled till they had reached a particular heart rate. At various points the researchers measured oxygen intake, blood pressure and glucose levels in the blood. The tests showed that not only did Red Bull not give them wings – the addition of the energy drink to alcohol made no noticeable difference to fitness.

Two years later the same researchers looked into the psychological effects of mixing alcohol with an energy drink. Compared to drinking just vodka, volunteers felt much fitter and sharper after drinking vodka and Red Bull. But tests showed that drinkers of just vodka and drinkers of vodka-Red

Bull scored equally badly on reaction speed and coordination tests. Breathalyser tests also showed no difference. Conclusion: adding energy drinks to spirits makes us reckless.

de Mello, M. T. & Ferreira, S. E. 'Does an energy drink modify the effects of alcohol in a maximal effort test?' in *Alcoholism: Clinical & Experimental Research* 28, 9 (2004), pp. 1401–1412

de Mello, M. T. & Ferreira, S. E. 'Effects of energy drink ingestion on alcohol intoxication' in *Alcoholism: Clinical and Experimental Research* 30, 4 (2006), pp. 598–605

DRINK ALCOHOL (AND LIVE LONGER)

Scientists have been arguing about the pros and cons of alcohol for years. Red wine is supposed to reduce the chances of a heart attack, but at the same time alcohol will increase the chance of certain types of cancer. What to do? To drink or not to drink?

When there are varying research results scientists often make a meta-study, where the data from various experiments is compared in the search for the ultimate answer. The Italian Catholic University of Campobasso looked at 34 studies into alcohol consumption and death figures. From the data, which covered over a million people, it was very clear that moderate drinkers (up to two glasses per day) lived longer than complete abstainers. But beware; if you drink more than these 'daily recommended doses' you enter the danger zone.

Is it relevant what type of alcohol you drink? It would appear so, given the results of 40 years of research conducted by the Wageningen University and the Ministry for Community Health and Environment in the Netherlands. They discovered that moderate wine drinkers lived on average a full two years longer than lovers of other alcoholic beverages, and nearly four years longer than teetotallers.

di Castelnuovo, A. & Costanzo, S. 'Alcohol dosing and total mortality in men and women: an updated meta-analysis of 34 prospective studies' in *Archives of Internal Medicine* 166 (2006), pp. 2437–2445

TALK ABOUT HOOLIGANS (AND WIN TRIVIAL PURSUIT)

SEVEN TIPS FOR YOUR SPARE TIME

GO BOWLING WITH STRANGERS

Well, it came as a big surprise to Tayyab Rashid of University of Pennsylvania. The psychologist signed up 85 students for a study in which he sent them bowling in the campus bowling alley. Some were given their own lane, others were four to a lane. Before and after playing the participants had to fill in an extensive questionnaire to establish how happy they were at that moment. The prediction was that bowling alone would make students less happy than bowling with others. That much turned out to be true.

And yet something strange was going on. Some groups were made up of four friends, but in other groups Rashid had put together people who did not know each other. Totally against expectations, it appeared that the bowlers who had shared the lane with strangers had a much better evening than the groups of friends. How is this possible? Rashid suspects that the tension involved in getting to know new people plays a role. Moreover, friends often carry old grudges. Groups of strangers start with a clean slate.

Rashid, T. 'Bowling alone or with others: who is happier?' Poster presented at the Fourth International Positive Psychology Summit, Washington DC (2005)

CHEAT (AND WIN EVERY 'GUESS THE NUMBER OF BEANS' COMPETITION)

No trip to the fair would be complete without having a go at 'Guess the Number of Beans'. Punters line up for their turn at divining the number of beans in a jar on a stall, in the hope of winning a trip to Morocco. But how to do it? To find the answer we need to go back over a hundred years.

In 1906 the British scientist Francis Galton, half-cousin of Charles Darwin, visited the West of England Fat Stock and Poultry Exhibition in Plymouth. Visitors were asked to guess the weight of an ox. For 6 pence anyone could have a go. Not just butchers, farmers and cattle dealers, but also ordinary members of the public took part. Afterwards Galton analysed the guesses of all 787 participants. He nearly fell off his chair in surprise. The average guess was 1,197 pounds. And what did the slaughtered ox actually weigh? 1,198 pounds. The group as a whole seemed extraordinarily clever.

Is that coincidence? Not in the slightest. This phenomenon has been shown often since Galton's publication in the scientific magazine *Nature*. The cause? Guesses that are too high are compensated for by guesses that are too low. So the extremes cancel each other out. The conclusion? If you need to guess the number of beans in a jar, you would do well to take a lot of friends with you and to average their guesses. No friends? Then secretly look over the shoulders of as many participants as possible and base your guess on theirs.

Galton, F. 'Vox populi' in Nature 75 (1907), pp. 450–451

SKIMMING STONES? GET THE ANGLE RIGHT!

In a bright yellow T-shirt and with his long hair under a baseball cap, American Kurt Steiner set the Guinness World Record for stone skimming in 2002. In Franklin, Pennsylvania, his stone skipped no less than 40 times over the water's surface.

What is the secret to success in this 'sport'? That's what Christophe Clanet of the University of Provence in Marseille asked himself. In his laboratory he sent aluminium stones of various weights, shapes and sizes over a body of water. He kept changing the speed at which the stone was catapulted, the speed of the stone's rotation and the angle at which it hit the water. With a high speed video camera Clanet filmed the skimming, and formulated the following advice for would-be stone-skimming champions:

1. Choose a flat stone.
2. Throw hard. World record holder Steiner probably threw at about 72 km per hour.
3. Allow the stone to rotate, that way it stays on course.
4. Make sure the stone hits the water at an angle of about 20 degrees. Then contact is as brief as possible and the least energy is lost.

Bocquet, L., Clanet, C. & Hersen, F. 'Secrets of successful stone-skipping' in *Nature* 247 (2004), p. 29

Bocquet, L., Clanet, C., Hersen, F. & Rosellini, L. 'Skipping stones' in *Journal of Fluid Mechanics* 543 (2005), pp. 137–146

TALK ABOUT HOOLIGANS (AND WIN TRIVIAL PURSUIT)

Surrounded by noisy supporters, a group of psychologists from Nijmegen University were walking to the football stadium. When one saw an empty beer can on the ground, he couldn't control himself and gave it a good kick. Why did he do that, his colleagues asked themselves. Had he unconsciously taken on the nearby hooligans' behaviour?

Researcher Ap Dijksterhuis decided to test this theory in the lab. He asked a number of his volunteers to write down the characteristics they thought belonged to the term 'football supporter'. A control group did nothing. Then everyone was given 42 questions from the game Trivial Pursuit. What transpired? The people who had thought about supporters scored on average less than the other group. An opposite effect was also noted by Dijksterhuis: thinking about professors beforehand resulted in higher scores on the Trivial Pursuit questions.

This phenomenon of unconscious influence is known in psychology as 'priming': thinking about older people makes you walk slower, thinking about hooligans impedes your intellect. According to Dijksterhuis it is not known if you can prime yourself by, for example, purposefully thinking really hard about professors before a game of Trivial Pursuit. But reducing the opposition's chances is possible: just strike up a conversation about hooligans.

Dijksterhuis, A. & van Knippenberg, A. 'The relation between perception and behaviour, or how to win a game of Trivial Pursuit' in *Journal of Personality and Social Psychology* 74 (1998), pp. 865–877

USE A LOT OF SUNSCREEN (OTHERWISE THE CREAM TAKES REVENGE)

It may come as a shock to learn that sunscreen, a product that is supposed to protect our skin, can itself harm our skin. What is going on here?

If skin molecules are exposed to sunlight, they absorb ultraviolet (UV) rays. During this process free oxygen radicals are created, which in time can cause all sorts of nastiness – from old age wrinkles to skin cancer. Luckily, sunscreens with a

UV-filter that reduces the number of UV-rays that penetrate the skin are available. But these creams also have a strange characteristic, as chemist Kerry Hanson of the sun-drenched University of California discovered. As soon as the UV-filters of the sunscreen are absorbed by the skin, they themselves make, under the influence of sunlight, free oxygen radicals. In this way the sunscreen adds to the natural creation of radicals. Should we stop slapping on the cream then? Definitely not. The effect only occurs when a little sunscreen is applied. So the advice is: during a day of sunning smear yourself regularly with a high factor sunscreen. And don't be stingy.

Bardeen, C. J., Gratton, E. & Hanson, K. M. 'Sunscreen enhancement of UV-induced reactive oxygen species in the skin' in *Free Radical Biology and Medicine* (2006)

TAKING A GROUP PHOTO? FIRST GRAB THE CALCULATOR

Yet another lovely family photograph with everyone smiling… except you've got your eyes closed. There's a strong chance that the photographer is not aware of Piers Barnes's calculations.

TALK ABOUT HOOLIGANS

This Australian physicist came up with a formula to calculate how many photos a photographer should take to be 99 per cent sure that everyone pictured would have their eyes open. First of all he established how often people blink when they are having their photo taken: about ten times a minute. Then he discovered that one blink lasted about a quarter of a second. The time to take a photo, in a well-lit space, he took to be 8 milliseconds. With fairly straightforward statistics he arrived at a nice, exponentially rising curve. But because most photographers like to keep things simple, he also came up with a rule of thumb: divide the size of the group by three if the light is good, and in half if the light is poor. The result of this calculation is the number of photos you must take. So, if 11 family members are posing on a sunny day on the crazy golf course? Take at least four photos. And if that same group of 11 want their photo taken with a beer in hand in a dimly lit restaurant? Then take at least six. Barnes's rule of thumb counts for groups of up to 20.

Barnes, P. & Svenson, N. 'Blink-free photos, guaranteed' in *Australasian Science* 10 (2006), p. 48

GET A DOG (AND MAKE FRIENDS)

A dog is good for so many things. In an experiment conducted by the University of Warwick in the UK a woman was given the company of a white Labrador for a week. The animal accompanied her on her usual daily activities: dropping the kids off, doing the shopping and working at the university. The dog, which had come from a blind dog school, was specially trained: the animal was not allowed to make contact with other humans itself; for example, by running up to people or by barking at them. The woman also followed strict social instructions – she was not allowed to greet first if she saw a familiar face. An unobtrusive observer followed the duo and tallied up all the social interactions, neatly divided into nods, smiles, short conversations, long conversations, etc.

  Result: with dog the woman had more contact than without. In particular, strangers made contact remarkably more often: three times in a week without the dog, but a massive 65 times in one week with the dog. The explanation is obvious:

people speak to each other more easily when there is a safe conversation topic.

Collis, G. M. & McNicholas, J. 'Dogs as catalysts for social interactions: robustness of the effect' in *British Journal of Psychology* 91 (2000), pp. 61–70

FIRST KISS? TILT YOUR HEAD TO THE RIGHT

FOURTEEN TIPS FOR A BETTER LOVE LIFE

DON'T WEAR GLASSES IF YOU ARE ON THE PULL

It must have been a wild night. Health Psychologist June McNicolas of the University of Warwick took a busload of short-sighted students to a prestigious London night club. The task she gave the volunteers was unambiguous: flirt, flirt, flirt! The 38 volunteers were chosen because they wear glasses or contact lenses in their day-to-day lives. For the experiment they were split into three groups: the first wore their usual form of vision correction; the second swapped their glasses for lenses; the third their lenses for glasses.

The students were asked to fill in a questionnaire after the event. The findings revealed that 85 per cent of those who had swapped glasses for lenses found that their confidence had thereby grown. The glasses wearers scored markedly worse; 80 per cent felt less attractive than with lenses.

According to researcher McNicolas this is not just imagination. No matter how trendy your glasses, it detracts from your most attractive body part: your eyes. The success rate of the flirting students speaks volumes. Compared to wearers of glasses, the contact lens wearers reported four times as many snogs.

McNicolas, J. Dept of Psychology, University of Warwick

WOMEN ONLY
DO IT WITHOUT
(BECAUSE SPERM MAKES YOU HAPPY)

Government campaigns have been urging us in recent years to practice safe sex. But guess what? Women run less chance of getting depressed if they do it without a condom. Gordon Gallup of the State University of New York asked 293 students how often their partners rolled on a condom when between the sheets. Using a questionnaire he assessed the ladies for potential symptoms of depression. These were the results:

1. Women who have sex without a condom are happier than those who do use one.
2. Women who don't have any sex at all are just as happy as women who do it with a condom.

3. Women who don't use condoms become unhappier the longer they don't have sex.

From this Gallup concluded that sperm must contain something that makes women happy. The researcher hasn't yet investigated if this 'trick' also works for oral or anal intake of sperm.

Burch, R. & Gallup, G. 'Does semen have antidepressant properties?' in *Archives of Sexual Behaviour* 31, 3 (2001), pp. 289–293

35.

MEN ONLY
ONLINE DATERS: EXAGGERATE YOUR INCOME

Age: 27
Favourite Dish: bubble and squeak
Hobbies: jogging and reading thrillers
Looking for: wild nights

So your online dating profile is complete. Now it's time to sit back and watch the responses flood in. But for some reason the number of people clicking on your profile isn't exactly shooting through the roof. Can you do anything to improve this? Definitely.

American economists at the University of Chicago investigated who was most successful in looking for a partner via the Internet. A dating site gave them an enormous amount of data. The researchers recorded which profiles 23,000 online daters viewed over three months, who they consequently made contact with and whether telephone numbers or email addresses were exchanged. It turned out that men with a salary of over $250,000 received 2.5 times as much attention as men with less than $50,000 on their pay-slips. For women, bragging about their income made no difference.

Some other telling conclusions of the research were:

- Men who claimed to be after a permanent relationship got on average more responses than others.
- If a man said he was after a casual affair he received on average 42 per cent less mail. However, women who showed they were up for a short-term adventure were more popular.
- At least 30 per cent of the daters had posted one or more photos on their profile. Men with photos got 1.5 times more responses than men without a photo, whereas women with photos got twice as many.

Ariely, D., Hitsch, G. & Hortascu, A. 'What makes you click? Mate preferences and matching outcomes in online dating', MIT Sloan Research paper 4603-06 (2006)

WOMEN ONLY
ALLOW HIM TO DO SCARY THINGS (AND HE'LL LIKE YOU MORE)

A wobbly footbridge, 70 m above a river – that would be a good place to conduct a research project, thought Don Dutton of the University of British Columbia. He asked a female interviewer to approach men between 18 and 35 who had just crossed the bridge. She told them she was busy with a project to analyse the effect of beautiful surroundings on creativity, and asked if they would mind writing a short story about the young woman on a photo she had with her. When the guinea pigs had completed their stories, she tore off a piece of paper and wrote her number on it, saying they could call her if they had further questions about her research.

In fact, the experiment had nothing to do with the effect of beautiful surroundings on creativity. This was a ruse to find out whether men get more aroused if they have just done something thrilling. Dutton therefore rated the sexual elements of the men's stories on a scale of 1 (no sexual content) to 5 (high sexual content). The results? Men definitely get aroused by thrills. Those who took part in the exercise when they had just crossed the bridge scored on average at least a point higher that the control group, which consisted of men who had had some ten minutes to recover from the bridge-crossing in an adjacent park. But that was not all. More of the men from the first group than from the control group rang the number the researcher had given out.

Did women get equally aroused by a male researcher after walking over a wobbly bridge? Dutton investigated this too. The humiliating answer: no.

Aron, A. P. & Dutton, D. G. 'Some evidence for heightened sexual attraction under conditions of high anxiety' in *Journal of Personality and Social Psychology* 30, 4 (1974), pp. 510–517

37.

NEVER BORROW AN INFLATABLE DOLL

Doctor Harold Moi and nurse Ellen Kleist, working in a local hospital in Nanortalik, Greenland, were very confused when the skipper of a trawler arrived for a consultation. The man seemed to have caught gonorrhoea a couple of weeks earlier. This was strange, because the man had been on a boat without women for three months. Had he had any homosexual contacts? He denied that was the case. But after some hesitation, the patient admitted that he once knocked on the engineer's door to alert him that there was a problem with the engine. When the engineer left to check it out, the skipper saw an inflatable doll in the man's bed. He could not resist the temptation, and entered the cabin for a secret love-making session with the doll. A couple of days later, the physical troubles started.

To find out whether the gonorrhoea had been transmitted by the doll, the engineer was sought and finally found. And yes, he also appeared to have the same disease. He admitted that, just before the skipper had knocked on his door, he had ejaculated in the plastic vagina. Moi and Kleist found this case so interesting that they decided to publish it: this became the first scientific description of the transmission of a disease through an inflatable doll. In 1996 they received an Ig Nobel prize for their publication, a prize for science that first makes you laugh, and then makes you think.

Kleist, E. & Moi, H. 'Transmission of gonorrhea through an inflatable doll' in *Genitourinary Medicine* 69, 4 (1993), p. 322

BEWARE OF THE CLITOFING (AND OTHER TOYS)

Despite the presence of eight dildos, including the memorably named Clitofing, Crystal Jellies Double Dong, Spectra Gels Anal Plug and Anneau d'érection, in the laboratory, it didn't get *that* exciting at the Dutch research organisation TNO. But they did make some important discoveries about what chemicals go into sex toys.

The researchers from Apeldoorn in the Netherlands cut the toys into one by 2 millimetre pieces. They then put the pieces through a chemical analysis. The burning question was: did they contain phthalates? These chemicals are used as plasticizers. They make plastic soft and flexible, but are harmful to humans: they can lead to disturbed hormone activity, liver and kidney problems and possible cancer.

As expected, most of the toys turned out to contain softeners. In seven out of eight they were present at high levels. Only the Cyber Pussy was completely plasticizer-free. This brought the researchers to the alarming conclusion that even sex with a toy is not completely safe. Perhaps people will have to put condoms over them to be sure?

Houtzager, M. M. G. & Peters, R. J. B. 'Determination of phthalates in sex toys' TNO Built Environment and Geosciences (2006), TR 2006/372

39.

IN THE MOOD FOR SEX? JUST ASK!

Four male and four female students, whose physical attributes ranged from 'slightly unattractive to moderately attractive', were given an unusual task. Researcher Russell Clark of Florida State University wanted to find out how direct you can be if you like someone. He sent the students onto campus with the task of approaching attractive strangers. They had to say: 'I have been noticing you around campus. I find you very attractive.' And then they had to pose one of the following questions:

1. Would you go out with me tonight?
2. Would you come over to my apartment tonight?
3. Would you go to bed with me tonight?

How did those asked react? Over half appeared ready for a date, and of those the women slightly more often then the men. But with the second and third questions there was a definite difference between the sexes. The women didn't want to come to the apartment and definitely not to bed with a strange man. And the men? They were very enthusiastic. At least three quarters of them were prepared to get under the sheets with a complete stranger. Some even answered: 'Why do we have to wait until tonight?' or 'I cannot tonight, but tomorrow would be fine.'

Type of request	Date	Apartment	Bed
Man asking	56%	6%	0%
Woman asking	50%	69%	75%

The social-biologist has an explanation for these differing reactions. Both men and women want to produce healthy children. To achieve this, a man only has to deposit his semen at the right moment in the right place. But a woman has to be more critical. She can only have a certain number of children in her life. Thus it is in her interest to choose the father carefully. That takes time. To jump into bed straight away would therefore not be sensible.

Clark III, R. D. & Hatfield, E. 'Gender differences in receptivity to sexual offers' in *Journal of Psychology and Human Sexuality* 2 (1989), pp. 39–55

TO MAXIMISE PLEASURE: DO IT TOGETHER

Sex with someone else is better than sex with yourself. To be more precise: four times better. This was the conclusion drawn by Stuart Brody and colleagues of the University of Paisley in Scotland when they re-analysed the figures from an earlier German study. For that project students were invited via an advert in the paper to come and make love in the laboratory. The area had been decked out for the occasion with a television showing erotic films. Some volunteers had sex with their own partners, others masturbated. Meanwhile, their blood was monitored. The interest was in the hormone prolactin that produces feelings of pleasure. What transpired? For both women and men, there was four times as much of this hormone present after an orgasm through penetration than after a masturbation-induced orgasm. Is that odd? From an evolutionary point of view it's understandable. We exist to procreate and keep the human race going. And so activities that help achieve this are rewarded.

Brody, S. & Krüger, T. H. C. 'The post-orgasmic prolactin increase following intercourse is greater than following masturbation and suggests greater satiety' in *Biological Psychology* 71 (2005), pp. 312–315

TILT YOUR HEAD TO THE RIGHT FOR THE FIRST KISS

It's been in the air for a while, but now it is really going to happen. The first kiss with your new flame, smack on the lips. There is just one practical problem: which way should you tilt your head to avoid bashing noses?

Go for the right, and you will be doing what 65 per cent of people do. This discovery was made by Onur Güntürkün of Ruhr University Bochum in Germany. The psychologist spied on 124 kisses in airports, train stations and other public spaces where many different nationalities can be found. Güntürkün's amorous score card was published just before St Valentine's Day in *Nature*, one of the world's most prestigious scientific magazines.

Furthermore, observations of young and unborn babies (through echograms of pregnant women) show that our tendency to tilt our heads to the right starts much earlier than the first kiss.

Güntürkün, O. 'Adult persistence of head-turning asymmetry' in *Nature* 421, 6924 (2003), p. 711

42.

MEN ONLY LOOK AT OTHER MEN (AND GET MORE POWERFUL SPERM)

The 52 heterosexual men who took part in Sarah Kilgallon's experiment were not allowed to have sex for two whole days. After their period of abstinence, the evolutionary biologist from the University of Western Australia gave each of these men a sealed envelope containing arousing material. The volunteers retreated to a private space and each opened their little gift. One half of the group received a sexually explicit photo of two men and one woman, the others a pornographic picture of three women. Next, the men had to get to work. While looking at their photos they deposited their sperm in a beaker, noted the time the ejaculation took, kept the precious goods warm under the arm or in a trouser pocket, and dropped it off within two hours at the lab. There Kilgallon compared the sperm quality of both groups.

The results? The gentlemen who had been stimulated by the photo of two men and one woman had more active sperm cells. Kilgallon explains this using 'sperm competition' theory: as soon as a man sees a competitor, he produces stronger sperm.

Kilgallon, S. G. & Simmons, L. W. 'Image content influences men's semen quality' in *Biology Letters* 1 (2005), pp. 253–255

WOMEN ONLY
DON'T MARRY A TEACHER

Does a man find his wife less attractive just after reading *Playboy Magazine*? Yes, he does, according to research by American psychologist Douglas Kenrich in 1989. Worse than that: after seeing a large quantity of hot babes men rate their entire relationship less highly.

How interesting, thought evolutionary psychologist Satoshi Kanazawa of Indiana University. What does this mean for men who come into contact with a lot of young women every day at work? Are they more likely to get divorced because all this female distraction influences their feelings about their marriage? Kanazawa studied the data of the General Social Survey, a database that includes profession and marital status for a large group of US citizens. He tested his hypothesis on male teachers, many of whom are surrounded by women in their most fertile years. The conclusion? Yes, male teachers get divorced more often than men in other professions. Women married to a high school teacher or university lecturer are especially at risk. Women married to a primary school teacher have nothing to fear.

Kanazawa, S. & Still, M. C. 'Teaching may be hazardous to your marriage' in *Evolution and Human Behaviour* 21 (2000), pp. 185–190

44.

SUPPORT EACH OTHER THROUGH THE GOOD TIMES

What are the signs of a good relationship? When two people get through the tough times together. That's right, isn't it? Yes, but according to the University of California there is a more important gauge: how partners behave towards each other when times are good.

Psychologists asked 79 heterosexual couples to tell each other of a recent positive event and of a recent negative one. Afterwards each wrote down how they had experienced their partner's reactions. For example, did it make them even happier in their relationship or not? The conversations were recorded on video and watched by a third party. As expected, the relationship was judged to be better by both the volunteers and the video-watchers the more enthusiastic and interested the partner's reaction to the positive story. Interestingly, the reaction to the negative event was less important: for men this didn't even have the slightest influence on their satisfaction with their relationship.

After two months the couples again answered questions about their love life. From this it was clear that a person's reaction on camera to their partner's positive stories was a better indication of how well the relationship was going than the reactions they had to each other's negative stories. Why is this? The researchers think that well-meant support for problems can have a contrary effect: it is in fact a recognition that there is something wrong, and a kind of supporter/victim relationship can develop. This danger is not present when reacting enthusiastically to a positive story. Conclusion: you can certainly comfort loved ones when they have it tough, but don't forget to celebrate their successes.

Gable, S. H., Gonzaga, G. C. & Strachman, A. 'Will you be there for me when things go right? Supportive responses to positive event disclosures' in *Journal of Personality and Social Psychology* 91, 5 (2006), pp. 904–917

45.

MEN ONLY
GET CIRCUMCISED

Condoms are the best defence against venereal diseases. But if you do have unsafe sex, there is less chance of getting infected with HIV if you are circumcised. In the South African Johannesburg region, at the instigation of the French Hopital d'Ambroise Pare, free circumcisions were advertised. After one and half years of recruitment this produced 3724 men. They first completed detailed questionnaires about their sex life. Did they use a condom? How many women had they slept with? The men were then split into two groups: one group was circumcised, the other not. Both groups had similar ages, societal status, condom use, etc. The men came back to be tested for venereal diseases and to complete questionnaires at regular intervals. The results after 6 months? Of the circumcised men, 20 per cent were HIV positive, compared to 49 per cent of those who had retained their foreskins. Inquiries in Kenya and Uganda delivered similar results and have been – just as the South African study – broken off in order to allow all the men to be circumcised. Why does circumcision help against HIV? Probably because the virus can stay alive longer in the humid area under the foreskin and can replicate itself better.

Auvert, B. & Taljaard, D. 'Randomized, controlled intervention trial of male circumcision for reduction of HIV infection risk: the ANRS 1265 trial' in *PLoS Medicine* 3, 5 (2005), p. 298

WOMEN ONLY
SNIFF BREASTFEEDING MOTHERS (AND GET IN THE MOOD FOR SEX)

The 47 young women who took part in a smell experiment at the University of Chicago had no idea which perfume they were dipping their noses in, but a certain number of them developed a marked increase in the desire for sex. So where did this lust-awakening perfume come from? Answer: from breastfeeding mothers.

In the interests of science, mothers placed cotton pads in their bras. These were then cut into pieces, which the childless volunteers smeared under their noses. They did this in the mornings, evenings, after sweating a lot and after a shower. In this way the experimental perfume was always present in the air.

The questionnaires that they were asked to complete daily revealed the following: after inhaling this earth mother smell single women had 17 per cent more sexual fantasies a day. Women with a partner had 24 per cent more. Why is the scent of milk, sweat and baby drool so exciting? There is a neat evolutionary explanation. If the circumstances are right for having a baby, it is useful for women to promote motherhood. Via their smell young mothers inspire other fertile women to reproduce.

McClintoch, M. K. & Spencer, N. A. 'Social chemosignals from breastfeeding women increase sexual motivation' in *Hormones and Behaviour* 46 (2004), pp. 362–370

BRUSH YOUR TEETH (AND REDUCE THE RISK OF A HEART ATTACK)

SIX TIPS FOR A HEALTHY BODY

EAT YOGHURT (AND HAVE FRESH BREATH)

How do you test if someone's breath is bad? By smelling it, you would have thought. But luckily for researchers at the Japanese Tsurumi University, there is a less unpleasant method: chromatography. This is a technique for the separation of a mixture by passing it through a medium in which the components move at different rates. It involves inserting a small amount of gas or liquid into test sample, of which the exact concentration can be established. During an experiment in which the Japanese researchers tested the breath of a group of volunteers, they paid special attention to sulphuration, which is known to create the smell of bad eggs. The root of the stink is to be found in the deep grooves at the back of the tongue. This is an oxygen-poor zone in which sulphate-producing bacteria thrive.

The volunteers stuck to strict rules regarding food, medicines and teeth brushing routines. After a zero-rating (the measuring of breath under normal circumstances) the volunteers ate 90 grams of sugar-free yoghurt twice a day for six weeks. This addition to their normal diet lowered the number of sulphate-producing bacteria significantly, with the direct consequence of fresher breath. This is probably thanks to the active bacteria in the yoghurt, with the poetic names *Lactobacillus bulgaricus* and *Streptococcus thermopilus*.

Hojo, K., Ohshima, T. & Yashima, A. 'Effects of yoghurt on the human oral microbiota and halitosis'. Eighty-third General Session of the International Association for Dental Research (2005), Baltimore, MD

MEN ONLY
DON'T WEAR YOUR TIE TIGHT

'Could you tighten your tie around you neck until it becomes a little uncomfortable?' That was the question doctors at the New York Eye and Ear Infirmary asked their male volunteers. After three minutes the ties were allowed to be loosened slightly. The doctors wanted to know to what extent a tight tie increased the risk of glaucoma, a generic term for eye problems where the humidity level of the eye is disturbed. Glaucoma can cause the pressure on the eye to increase to such an extent that the blood flow to the eye nerve is blocked, with the possible consequence that the nerve is damaged and peripheral vision lost. In extreme cases this tunnel vision can deteriorate into blindness. The experiment showed that a tight tie significantly increases the pressure on the eyeball, by a good 15 per cent for the average man with healthy eyes. So if you care about your eyes, you better not pull your classic knot or double Windsor too hard.

Teng, C. 'Effect of a tight necktie on intraocular pressure' in *British Journal of Ophthalmology* 87 (2003), pp. 946–948

BRUSH YOUR TEETH
(AND REDUCE THE RISK OF A HEART ATTACK)

We already knew that maintaining a high level of oral hygiene is good for your teeth. But at Columbia University in America they have discovered that brushing your teeth also helps reduce the risk of cardiovascular disease.

Researchers analysed the bacteria in the mouths of 657 volunteers who had never had a stroke or heart attack. They also noted the thickness of the carotid artery, which carries blood from the heart to the brain. The width of this artery is a good indication of arteriosclerosis, when cholesterol or other substances nestle in the walls of the carotid artery. When this happens the blood flow can be affected and even stop.

So did the researchers find a connection between oral hygiene and healthy arteries? Indeed they did. The more bacteria a gum infection creates, the thicker the carotid artery. And that means greater risk of a stroke or heart attack. The researchers suspect that the bacteria travel through the body from the mouth via the blood and cause infections which close the arteries. How to avoid all this? Simple: through good brushing and flossing.

Demmer, T. T. & Desvarieux, M. 'Periodontal microbiota and carotid intima-media thickness: the oral infections and vascular disease epidemiology study' in *Circulation* 111, 5 (2005), pp. 576–582

DON'T GET A NIPPLE PIERCING (AND DEFINITELY NOT A PRINCE ALBERT)

Don't go and pierce your private parts just yet. First read the warning from Carol Caliendo, Professor of Health Studies at Carlow University in Pittsburgh. She found 146 people across America with intimate piercings (navel, genital or both) prepared to fill in a questionnaire. Although most metal wearers were happy with their piercings, they admitted to a remarkable number of health problems. Sixty-six per cent of people with a nipple piercing suffered from skin irritations or infections. Of the men with a penis piercing as many as 39 per cent complained of urinary problems. This was particularly true for those owners of the so-called 'Prince Albert', named after Queen Victoria's husband, who, persistent rumour has it, decorated his royal member with metal. With a Prince Albert a hole is drilled through the end of the urinary tract, allowing a ring to be inserted and protrude at the bottom end of the penis head. Penis piercings also often lead to torn condoms: one in five of the volunteers had experienced this.

Armstrong, M. L., Caliendo, C. & Roberts, A. E. 'Self-reported characteristics of women and men with intimate body piercings' in *Journal of Advanced Nursing* 49, 5 (2005), pp. 474–484

WASH WITHOUT SOAP ONCE IN A WHILE

If your feet smell, wash them with soap! It seems perfectly good advice, but it isn't always. Too much soap disturbs the natural pH balance of the skin, and therefore just creates more odour.

A research project that examined 1,270 Kenyan prostitutes showed that soap has other negative effects. An international team of researchers under the leadership of the University of Washington asked the women if they cleaned their vaginas. The majority did: 23 per cent washed with water; 71 per cent added soap or some other cleansing product; and a small minority, 6 per cent, didn't clean their vaginas at all.

Over a period of ten years, these women were tested monthly for HIV. The results of the survey suggested that the women who didn't clean their intimate parts had the least chance of catching the deadly virus. In terms of this group, washing out with water increased the risk of contracting HIV three-fold. For those who used soap it was even worse: nearly four times higher risk of infection. This is also due to the affect on the pH of the skin.

So the advice would appear to be: go easy on the soap when showering, or use a special soap with the same pH as the skin (indicated on the label as a pH level of about 5.5).

Hassan, W. M., Lavreys, L. & McClelland, R. S. 'Vaginal washing and increased risk of HIV-1 acquisition among African women: a 10-year prospective study' in *AIDS* 20, 9 (2006), pp. 1347–1348

SLEEP IN (AND LOSE WEIGHT)

A few kilos too heavy? Don't immediately start dieting – wait until you hear how sleeping a bit longer can affect your weight.

Researchers at the Clinical Research Centre of the University of Chicago asked 12 healthy, non-smoking young men with normal posture to spend two nights in the hospital on two separate occasions. The first time they were allowed to sleep for four hours a night (from 1 a.m. to 5 a.m.), the second time they were allowed to sleep for ten hours a night (from 10 p.m. to 8 a.m.). Although the men had exactly the same amount to eat, they appeared to be much more hungry in the morning after a short sleep than after a long sleep. They particularly wanted energy-rich food like ice cream, chips and pasta.

Why is this? Blood tests showed that lack of sleep had confused the production of two hormones in the men. The short sleepers had more ghrelin, a hormone which stimulates hunger. The level of the hormone leptin, which sends signals that we have had enough to eat, was lower after a short night.

But do these hormones really make us fat? The results of 15 years of research into sleep by the University of Wisconsin were published in 2004, showing that people who sleep less than eight hours a night are on average fatter.

Spiegel, K. & Tasali, E. 'Brief communication: sleep curtailment in healthy young men is associated with decreased leptin levels, elevated ghrelin levels, and increased hunger and appetite' in *Annals of Internal Medicine* 141 (2004), pp. 846–850

Lin, L. & Taheri, S. 'Short sleep duration is associated with reduced leptin, elevated ghrelin and increased body mass index' in *PloS Medicine* 1, 3 (2004), p. 62

SPREAD LEMON SCENT (AND EVERYBODY WILL START CLEANING)

FOUR TIPS FOR IN AND AROUND THE HOUSE

DON'T GET A CLEANER
(AND REDUCE THE RISK OF CANCER)

No time to clean the windows, wipe the fridge or mop the kitchen floor? You might be tempted to hire a cleaner – but that is not without its dangers, a recent study shows.

Under the auspices of the German Institute for Food Research Potsdam-Rehbrucke, several researchers from throughout Europe took part in a study. During many years they followed 218,169 women from nine countries. The women, between 20 and 80 years old, had to say how much exercise they got. Did they do office work or heavy manual labour? Were they active in household chores and did they like working in the garden? How often did they take part in any sporting activity? Then the researchers noted which women, during the period of research, got breast cancer.

After in-depth analysis of the figures it appeared that housework, like cleaning, washing, cooking and looking after children, reduced the risk of breast cancer. The researchers suspect that regular small exertions have more effect on the prevention of breast cancer than more intensive, less frequent exertions, like sport.

Friedenreich, C. & Lahmann, P. H. 'Physical activity and breast cancer risk: the European prospective investigation into cancer and nutrition' in *Cancer Epidemiology Biomarkers and Prevention* 16, 1 (2007), pp. 36–42

STORE YOUR VALUABLES IN THE CHILDREN'S BEDROOMS

How do burglars go about their business? Claire Nee and Amy Meenaghan from the University of Portsmouth decided to visit a jail and ask the criminals themselves. During the selection process, the psychologists were very strict: only those who had committed at least 20 burglaries in the last three years were included. Fifty burglars, varying in age from 21 to 60, agreed to be interviewed. Nine out of ten said that they worked according to a certain pattern during a burglary. Most of them started off in the adult bedrooms, continued into the living room, the dining room and the study, finally ending their quest in the kitchen. 'People leave things in the same basic locations… could have done it with my eyes shut,' one of the interviewees told the researchers. Only one question remains: what were the rooms that they left untouched? The answer: the children's rooms. There, hardly anything of any value was ever found.

Nee, C. & Meenaghan, A. 'Expert decision making in burglars' in *British Journal of Criminology* 46 (2006), pp. 935–949

SPREAD LEMON SCENT (AND EVERYBODY WILL START CLEANING)

It's always the same; you cleaned up last week and now the sitting room/office/car is a mess again already. Here is some advice from social psychologist Rob Holland: spread lemon scent, and everyone will keep the place much tidier.

For a reward of one euro, 22 students from Radboud University in Nijmegen, The Netherlands, filled in a questionnaire. Half were put in a room in which the typical lemon smell of cleaning material hung in the air, the other half in a room with no obvious odour. After filling in the questionnaire the volunteers ate a biscuit in another room. Without them being aware of it a camera registered their every hand movement. What did careful analysis of the data show? Those who had been exposed to the lemon scent wiped crumbs from the table three times more often than those who had been in the neutral smelling room.

What made the students react in this way? Everything in our environment, from background noises to advertising boards, influences our behaviour. This is true of smells too. When we smell freshly baked bread, we immediately feel hungry. We associate the smell of cleaning products with gleaming bathrooms and sparkling kitchens. This is why a lemon scent spurs us to start cleaning. A small amount is enough; of the volunteers, only one had consciously smelled lemon in the air.

Aarts, H., Hendriks, M. & Holland, R. W. 'Smells like clean spirit: non-conscious effects of scent on cognition and behaviour' in *Psychological Science* 16 (2005), pp. 689–693

THROW YOUR SHOWER CURTAIN OUT

If you are not careful, you can catch nasty things in the shower. Scott Kelley of San Diego University and three colleagues from other universities decided to take their shower curtains to the lab for analysis. All the curtains had been in use for a minimum of six months without having been properly washed. Analysis of the red and white fluff and pink, slimy layers brought the researchers to a halt. Hundreds of millions of bacteria had taken up residence!

Sphingomonas and *Methylobacterium* appeared to especially like the bathroom. These bacteria are not entirely friendly; in fact, 80 per cent of the types found are known to be unfriendly. On the whole they don't harm healthy people, but they can cause problems for less robust people, like patients suffering from illnesses such as Aids or cancer.

So what should we do? Researcher Norman Pace advises that we throw the shower curtain in the washing machine every couple of weeks.

Kelley, S. 'Molecular analysis of shower curtain biofilm microbes' in *Applied and Environmental Microbiology* 70, 7 (2004), pp. 4187–4192

DUNK YOUR BISCUIT HORIZONTALLY, NOT VERTICALLY

TWELVE TIPS FOR EATING AND DRINKING

57.

SHAKE YOUR MILKSHAKE LONGER (AND GET THINNER)

Eat bigger portions and lose weight. It seems counter-intuitive but in some cases it is true. American Professor of Food Science Barbara Rolls of Penn's State College asked 28 slender men to come and have breakfast, lunch and dinner once a week for four weeks in her lab. Half an hour before lunch a delicious strawberry milkshake was served. The gentlemen did not know that the professor had left some milkshakes longer in the blender than others. All had the same number of calories, but some were larger because they contained more air. Did this affect the volunteers' behaviour? Indeed it did. The men who drank the airiest (and to look at bigger) milkshakes ate 12 per cent less calories during lunch than those who had had the smaller looking shakes. Nor did this extra air cause more hunger or tiredness later on.

We can therefore manipulate our feeling of 'being full' by pumping more air into our food. And air is not the only weapon to increase the volume of food while keeping the number of calories the same. Water works even better, and has the virtue of causing fewer burps. An example: 100 kilocalories of raisins fill a quarter of a coffee cup, while a 100 kilocalories of grapes fill nearly two coffee cups. If you chose the latter, you feel full sooner, stop eating and stay just that bit thinner.

Bell, E. A., Rolls, B. J. & Waugh, B. A. 'Increasing the volume of a good by incorporating air effects satiety in men' in *American Journal of Clinical Nutrition* 72, 2 (2000), pp. 361–368

DRINK COKE THROUGH A STRAW

Tell me how you drink and I'll predict how many cavities you will have. That is Mohamed Bassiouny's message, Professor of Dentistry at Temple University, Philadelphia. In a scientific article in *General Dentistry* he describes the mouths of two teenagers. The first belongs to an 18-year-old boy who consumes more than 2 litres of soft drink a day. He drinks it straight out of the bottle. The consequences are clear for all to see: his teeth are crumbling, they are stained brown and black and his breath reeks. To hide all this he presses his lips tightly together when he doesn't need to speak. The second mouth Bassiouny reports on belongs to a 16-year-old girl. In recent years she has consumed nearly 1.5 litres of soft drinks a day through a straw that ends just past her lips, in front of her teeth. Her teeth are also heavily affected, but in different places. Where the boy has especially damaged his molars at the back with the acid and sugar in the drinks, the girl has, by using the straw, damaged her front teeth. So how should you drink? If you can't resist the sweet stuff, Bassiouny recommends you drink it through a straw that goes deep into your mouth, beyond the teeth. Then contact between the teeth and the liquid is minimal.

Bassiouny, M. A. & Yang, J. 'Influence of drinking patterns of carbonated beverages on dental erosion' in *General Dentistry Journal* 53 (2005), pp. 205–210

COOK CHIPS IN A MICROWAVE

Frying chips? Put them into the microwave briefly before throwing them in the deep fat fryer. By pre-cooking them in this way they need less time in the fat, and this in turn means there will be less acrylamide, a cancerous substance, on the chips.

This is the result of research at the Turkish University of Mersin. The researchers bought sunflower oil and potatoes at a local market. They cut the peeled potatoes into chips of 7 centimetres. Then some of the chips went into the microwave before frying, whilst others were not pre-cooked. They were fried at different temperatures until they were ready. The amount of acrylamide in the pre-cooked chips was up to 60 per cent less than in those chips that had not been in the microwave. Frying at 150 °C rather than 190 °C also appeared to be a healthier option.

Ekiz, H. I., Erdogdu, S. B., Gökmen, V., Palazoglu, T. K. & Senyuua, H. Z. 'Reduction of acrylamide formation in French fries by microwave pre-cooking of potato strips' in *Journal of the Science of Food and Agriculture* 87, 1 (2007), pp. 133–137

EAT OFF THE TOILET SEAT INSTEAD OF OFF THE CHOPPING BOARD

Yummy, a big oven pizza. Because of its size we usually put it on a wooden chopping board to slice it up. But look out: it is more hygienic to eat if off the toilet seat. Pat Ruskin of the University of Arizona discovered that three times less bacteria live there. Even faecal molecules are present in greater concentration on the chopping board than on the toilet seat.

But how is this possible? Toilet seats are too smooth and dry to keep a great number of bacteria alive. The wood of a chopping board is full of grooves and even after thorough drying stays feeling damp. The faecal molecules that come into contact with the chopping board via dirty fingers can happily start colonies.

Gerba, C., Orosz-Coughlin, P. & Rusin, P. 'Reduction of faecal coliform, coliform and heterotrophic plate count bacteria in the household kitchen and bathroom by disinfection with hypochlorite cleaners' in *Journal of Applied Microbiology* 85 (1998), p. 819

CUT THE CABBAGE (AND KEEP IT FOR TWO DAYS)

It's a familiar dilemma in the vegetable section of the supermarket. Do you go for the whole cabbage or for the pre-chopped greens in a bag? Those who tend to go for the lazy option can now stop feeling guilty. Research from Wageningen University in the Netherlands shows that vegetables like cabbage, broccoli and sprouts get healthier if they are cut.

The researchers cut varieties of cabbage into small blocks of about one square centimetre. Portions of 100 grams were then stored. After two days the amount of glucosinolates had doubled in some varieties. And that is not bad news. Glucosinolates are sulphur-containing substances which boost the immune system and protect against geriatric illnesses.

Researcher Matthijs Dekker suspects that the cabbage mistakes the knife for an insect. In nature, when eaten by insects, cabbage generates more glucosinolates to protect itself. In this experiment, the cabbage had the same reaction when chopped with a knife.

Dekker, M., Jongen, W. M. F. & Verkerk, R. 'Post-harvest increase of indolyl glucosinolates in response to chopping and storage of Brassica vegetables' in *Journal of the Science of Food and Agriculture* 81 (2001), pp. 853–958

62.

BAD WINE? SERVE STILTON!

Wine and cheese: a long-celebrated food marriage made in heaven. But were these gastronomic partners really made for each other?

That is what Berenice Madrigal-Galan of the University of California asked herself. She bought eight kinds of red wine and eight different cheeses. First she asked 11 trained wine-tasters to judge the wine. When that had been done, the cheese was brought to the table. The cheese cubes (of exactly 5 grams each) were presented with toothpicks in plastic cups. She asked the tasters to put a whole cube into their mouths, to chew well and to then taste a sip of wine. They then had to judge the wine in terms of oak, mushrooms, mint and vanilla.

The result? With each cheese, it became harder to detect the flavours in the wine. This probably happens because the cheese covers the inside of the mouth with a layer of fat, making it more difficult for the wine to reach the taste buds. Blue cheeses like stilton and gorgonzola had the strongest effect.

Heymann, H. & Madrigal-Galan, B. 'Sensory effects of consuming cheese prior to evaluating red wine flavour' in *American Journal of Enology and Viticulture* 57, 1 (2006), pp. 12–22

USE OLIVE OIL!

When she was 85 she took up fencing; at 100 she was still riding her bicycle; and she only died when she was 122. The French woman Jeanne Calment (1875–1997) is the oldest woman who ever lived. Her secret? She loved chocolate, port and lots of olive oil. The latter could have been crucial, because it is possible that the liberal use of olive oil in southern Europe makes instances of certain cancers less frequent there than in northern Europe.

 Anja Machowetz of the German Institute of Human Nutrition in Potsdam-Rehbruecke and Henrik Poulsen of the Rigshospitalet in Copenhagen decided to look into the effect of olive oil in more depth. They asked 182 men from various

European countries to replace a portion of the fat they use every day with 25 ml Spanish olive oil specially prepared for the experiment. The olive oil that they might usually use they had to put aside. During the three-week duration of the experiment the researchers took urine samples. They looked for the waste products that are created when there is damage to genetic material. Accumulation of such damage to DNA can give rise to cancer.

And what transpired? After the olive oil diet far fewer harmful waste products were found in the urine. The differences are comparable to earlier research, where the researchers looked into the waste products in urine of people who had stopped smoking.

Gruendel, S., Machowetz, A. & Poulsen, H. E. 'Effect of olive oils on biomarkers of oxidative DNA stress in Northern and Southern Europeans' in *The FASEB Journal* 21, 1 (2007), pp. 45–52

64.

EAT SPINACH WITH YOUR STEAK

Mmm… a nice bit of steak. But beware: red meat increases the risk of bowel cancer. Through research on rats Johan de Vogel of Wageningen University discovered that the culprit was the substance heme iron. This molecule is in pretty much all meat: in beef the most, followed by pork and chicken. When De Vogel fed his rats heme iron it led to disruption in the way in which the cells in the intestine wall divided and died. By consequence, the risk of colon polyps and cancer increased.

Luckily, there is an antidote: thanks to the chlorophyll in the leaves, green vegetables can cancel out the harmful effects of red meat. The chlorophyll molecule fits like a piece in a puzzle onto the heme iron molecule, preventing it from disturbing the intestine wall. With the test rats this worked perfectly, and De Vogel thinks the effect can work the same for humans. From his lab animal research he calculated how much greenery a human would have to eat with a steak of 150 g in order to counteract the harm done by the red meat. Broccoli, sprouts or cabbage are clearly only suitable for those with big appetites:

Vegetable	Amount to be eaten with a 150 g steak
Spinach	75 g
Endive	75 g
Lettuce	100 g
Broccoli	750 g
Sprouts	950 g
Cabbage	3,500 g

de Vogel, J. 'Green Vegetables and Colon Cancer: the Mechanism of a Protective Effect by Chlorophyll' (2006), PhD Wageningen University

DUNK YOUR BISCUIT HORIZONTALLY, NOT VERTICALLY

Inexperienced dunkers often end up with a murky cup of tea with spongy biscuit remains in the bottom. That's because the sugar in the biscuit dissolves in the hot liquid and the fat melts. These are the substances that glue the rest of the ingredients together. On top of that, the starch crumbs soak up the moisture like a sponge. The result: the now soft biscuit will succumb under its own weight and break.

The British physicist Len Fisher of University of Bristol decided to solve this problem once and for all. In order to approach dunking with scientific precision, he put a biscuit under the microscope. He noticed that biscuits are full of holes connected to each other by little hollow channels. He then had the idea of dredging up a theory from 1921: the Washburn equation. This describes the speed at which porous material sucks up certain liquids. Using this formula you can calculate easily how fast toilet paper will suck up a pool of ink, but according to Fisher the same principle should also be applicable to biscuits in a cup of tea. The researcher dunked, calculated and finally arrived at the optimum dunking technique. The trick is as simple as it is brilliant: hold the biscuit as horizontally as possible, like a raft on a river. This way the top stays dry, while the underside sucks up liquid. The biscuit keeps its strength longer this way. Fisher calculated that this method allowed him to dunk four times longer than before.

Fisher, L. 'Physics takes the biscuit' in *Nature* 397 (1999), p. 469

Fisher, L. *How to Dunk a Donut* (2002, Phoenix Popular Science).

66.

SNIFF THROUGH YOUR RIGHT NOSTRIL (AND EVERYTHING SMELLS NICER)

If you really want to enjoy the smell of peppermint or vanilla, you should squeeze your left nostril shut and sniff the aroma only through your right nostril.

This is the striking conclusion reached by Larry Cahill of the University of California on the basis of the following experiment. Thirty-two volunteers were served up eight different smells, ranging from coconut to lemon. The perfume drops were kept in sealed beakers, so that the colour of the liquid would not give the game away. The volunteers sniffed each beaker for two seconds, while keeping one nostril closed. On a score card they rated each smell. After a one minute pause the researchers presented the next smell.

A week later the whole experiment was repeated but with one difference: now the volunteers had to sniff with the other nostril. Did this change anything? Absolutely, because sniffing with the right nostril gave the most pleasure, while the naming of perfumes was more successful using the left nostril.

The explanation for this is that our nostrils send all the information they gather through smelling to their corresponding side of the brain. The right side of the brain has the most to do with the processing of emotions, and therefore does the most enjoying of scents. The left side of the brain has more to do with language formation, and therefore we are better able to attach the words 'aniseed' and 'caramel' to their corresponding smells when we sense them through our left nostrils.

Herz, R. S. & McCahill, C. 'Hemispheric lateralisation in the processing of odor pleasantness versus odor names' in *Chemical Senses* 24 (1999), pp. 691–695

67.

DON'T GRILL AUBERGINES (AND SPARE YOUR TEETH)

Grilled vegetables may be just as bad for your teeth as Coke or orangeade. This is what dental expert Graham Chadwick discovered in the kitchen of the University of Dundee, Scotland. He tied on his apron and made two pans of ratatouille. In each pan he put exactly the same ingredients:

50 g onions
41 g green pepper
60 g aubergine
27 g courgette
58 g tomatoes
32 g red peppers
15 ml olive oil (extra virgin)

For the first pan he followed the traditional stewing method: he baked the vegetables, added water and left the lot to cook through. For the second ratatouille he grilled the vegetables. Chadwick then pureed both concoctions in a blender and

111

slipped into the lab to investigate the pH levels. The result? The stewed ratatouille was less acidic and therefore less harmful for the teeth. Chadwick also investigated which vegetables do the most harm. We should probably avoid grilling aubergines, courgettes and green peppers in the future.

Chadwick, R. G. 'The effect of cooking method upon the titratable acidity of a popular vegetarian dish: scope for reducing its erosive potential?' in *European Journal of Prosthodontics and Restorative Dentistry* 14 (2006), pp. 28–31

BECOME A VEGETARIAN (AND SAVE THE PLANET)

Factories and cars may emit the largest quantity of greenhouse gases but cattle make their contribution. Methane, the most important greenhouse gas after carbon dioxide, is excreted from the stomachs of these ruminants and is also given off from piles of their dung. The contribution of cattle emissions to global warming is, according to the University of Chicago, significant. Can we do anything about it?

Yes – we can stop eating meat. Research shows that this has a surprisingly big effect. The less meat eaters there are, the less demand for meat there will be – and therefore the fewer methane-releasing cows on the planet. A meat eater who becomes vegetarian contributes just as much to the lowering of greenhouse gasses as a driver who exchanges his gas guzzling SUV for a medium range car.

Eshel, G. & Martin, P. 'Diet, energy and global warming' in *Earth Interactions* 10 (2006), pp. 1–17

THINK ABOUT WEIGHT LIFTING (AND GET IN SHAPE)

SIX TIPS FOR SPORTS FANS

WRESTLE, BOX AND PLAY FOOTBALL IN RED

Does the colour of your strip affect the result of the match? That is what anthropologists at the University of Durham asked themselves. They watched carefully the matches of the 2004 Olympic Games in Athens. They concentrated on boxing, tae kwon do, Greek-Roman wrestling and freestyle wrestling. For these matches the participants are randomly given a red or a blue costume. Curiously, participants in red won all three sports more often than those wearing blue.

But what about team sports? To find the answer to this question, the researchers watched the 2004 European football cup. They looked at the results of five teams with a red strip that had also played wearing another colour in the same competition. What did they find? In their 'red' matches all five teams performed better, especially where the number of goals were concerned. How can this be? Animals associate red with masculinity, anger, aggression and testosterone, and so the researchers suspected that humans also react strongly to this colour. Their conclusion was that a red strip unconsciously fills the opposing team with angst and therefore affects their performance negatively.

Barton, R. A. & Hill, R. A. 'Red enhances human performance in contests' in *Nature* 435 (2005), p. 293

70.

SCORED A GOAL? CELEBRATE WITH RESTRAINT

If you score during a sports match, it's better to stay calm. Turkish research shows that one in twenty injuries occur during cheering. Over two seasons, 152 professional players had appointments with sports doctor Bulent Zeren in Izmir. To his surprise nine of them had received their injuries during the explosion of joy after a goal. Exactly what had happened?

To get to the bottom of it he decided to interview the unhappy celebrators again. It turned out that three players had been squashed by a heap of enthusiastic team members. This had led to a broken collarbone, a bruised rib and back pain. Five others were injured while sliding on their back, stomach or knees over the grass after scoring. One player, the sports doctor wrote, had fractured his ankle in this way. With his legs out before him he had slid further on the wet pitch than he had anticipated. He hit an advertising board and was out of action for 20 weeks. It's worth noting that no celebration injuries were received while doing a somersault or a back flip.

Öztekin, H. H. & Zeren, B 'Score-celebration injuries among soccer players' in *The American Journal of Sports Medicine* 33, 8 (2005), pp. 1237–1240

THINK ABOUT WEIGHT LIFTING (AND GET IN SHAPE)

Good news for the lazy: if you want to get in shape, you don't have to visit the gym. Cleveland Clinic Foundation in Ohio recruited volunteers for an unusual training program. First, the participants demonstrated the strength in their little finger with the help of specially developed fitness equipment. After this trial the volunteers followed a twelve-week program, during which they did little finger exercises for 15 minutes every day. However, they were only allowed to *think* about the exercises, and stay completely still. At the end of the training program the participants demonstrated their progress in an ultimate weight-lifting test. The results were staggering.

Participants who had only trained their little finger mentally gained an increase in little finger strength of 35 per cent. How is that possible? Our physical powers do not only come from our muscles, but also from our brain. Muscles are triggered by nerve cells called motor neurons, and you can train them by just thinking of your body's movements.

Liu, J. Z., Ranganathan, V. K. & Siemionow, V. 'From mental power to muscle power: gaining strength by using the mind' in *Neuropsychologia* 42 (2004), pp. 944–956

72.

DON'T BOTHER WITH SPORTS MASSAGE

How lovely – a nice massage after sport. But does it also help repair the muscles? No, says University College Northampton, UK. Eight amateur boxers each hit a special punch ball which measured impact 400 times. Immediately afterwards, half of them received a 20 minute sports massage, while the other half just had a rest without massage. At various intervals blood samples were taken through little finger punctures to measure, amongst other things, lactic acid levels. This substance is a good indication of muscle tiredness. It is assumed that a massage increases the blood circulation and so aids the absorption of lactic acid. But the blood analyses didn't show any of this.

Did the massaged boxers have more energy during their next boxing session an hour later? No, their punches were comparable to those of their colleagues who had not had a massage. A similar Scottish experiment, this time with cyclists measured on an ergometer, also showed massages to have no effect.

Hemmings, B. & Smith, M. 'Effects of massage on physiological restoration, perceived recovery, and repeated sports performance' in *British Journal of Sports Medicine* 34 (2000), pp. 109–114

Robertson, A. & Watt, J. M. 'Effects of leg massage on recovery from high intensity cycling exercise' in *British Journal of Sports Medicine* 38 (2004), pp. 173–176

73.

GO TO THE STADIUM
(AND THE REF WILL BE ON YOUR SIDE)

Referees are expected to be unbiased. Still, it seems they let themselves be influenced by the shouting from the crowds, says Thomas Dohmen of the German research centre IZA.

From analysis of 3,591 Bundesliga matches he discovered that referees decide in favour of the home team remarkably often: If the home team is one down at the end of regular play then the referee allows another 20 seconds of play; 20 more than if the home team is ahead by a goal. They are also more likely to give the home team a dubious goal or penalty.

But is this because of the jeers and cheers of the crowd in the stadium? Definitely, because referees particularly whistled in the home team's favour when the opposition had little support in the stadium.

What's more, the positioning of the stands plays a role. Are the supporters sitting close to the pitch? Then the referee is more inclined to decide in favour of the home team than if there is more distance; for example, if there is an athletics

121

track around the pitch. So the closer the crowd, the more the ref lets himself be influenced.

Dohmen, T. J. 'Social pressure influences decisions of individuals: evidence from the behaviour of football referees' IZA Discussion Paper 1595 (2005)

DON'T LET YOUR CHILDREN TAKE UP BOXING

Even highly educated pedagogues will sometimes admit it: 'Problem children can learn to control their anger outbursts through combat sports such as karate or boxing.' Nonsense, so research by Inger Endresen and Dan Olweus from Bergen University in Norway tells us. For three years they gave secondary school children a list of questions to answer. The volunteers, a total of 5,171 boys of 11 to 15 years old, noted down which sports they did and how many times a week they trained. The boys were also asked about their behaviour outside of sport. For example: 'During the last year did you start a fight with a class mate?' Or: 'Have you, during the last year, used a weapon in a fight, like a knife, a stick or something else?'

The inquiry was repeated one and two years later, so that the researchers could look carefully at the consequences of taking part in combat sports. The unambiguous conclusion: a visit to the dojo or the boxing ring doesn't work as a release at all, it makes teenagers more asocial and violent.

Endresen, I. M. & Olweus, D. 'Participation in power sports and antisocial involvement in preadolescent and adolescent boys' in *Journal of Child Psychology and Psychiatry* 46, 5 (2005), pp. 468–478

DON'T LOOK AT BEAUTIFUL WOMEN (IF YOU STILL HAVE THINKING TO DO)

FIVE TIPS FOR SMALL AND LARGE PURCHASES

75.

NEVER TRUST AN ESTATE AGENT

Estate agents know more about the housing market than home owners. That's why you employ them. But is that really such a sensible thing to do? Perhaps not. Research by Steven Levitt of the University of Chicago shows that estate agents don't always do their best for their clients.

The researcher studied the details of nearly a hundred thousand houses sold between 1992 and 2002 in Cook County in the American state of Illinois. In 3,330 cases the sale involved the private home of the estate agent. And that made a big difference. The agents let their own houses stay on the market for an average of ten days longer. The selling price was thereby an average of 3.7 per cent higher than the price of a similar house owned by someone else. Why is this? According to Levitt, the incentive for estate agents to drive up the price of their clients' houses is small. For example, say an offer of 200,000 pounds is made on a house. The agent gets 2 per cent of the selling price for his or her services: in this case 4,000 pounds. For the person who is selling the house it could be interesting to wait a little longer. Maybe someone will offer 10,000 pounds more. But that doesn't do much for the agent – those 10,000 extra pounds for the client mean just 200 pounds more for him or her. So what does the agent do? Advises the seller to accept the offer, of course: then he or she can move on to the next house.

Levitt, S. D. & Syverson, C. 'Market distortions when agents are better informed: the value of information in real estate' in NBER Working Papers Series, Working Paper 11053 (2005)

SLEEP ON IT FOR A NIGHT

Thinking about selling your house? Changing your job? Don't decide straight away; sleep on it for a night.

Ap Dijksterhuis of Amsterdam University in the Netherlands showed his volunteers four imaginary cars: the Hatsdun, the Kaiwa, the Dasuka and the Nabusi. Every car had four characteristics, positive and negative. One of the cars was obviously the best. After four minutes the participants had to say which car they would chose. Half the group could take this time to think hard about the cars. The other half had to solve puzzles. Who chose the best car most often? The people who had thought about it seriously.

But then Dijksterhuis changed the experiment. Now the cars had 12 rather than four distinguishing features. In this experiment the people who had been solving puzzles chose the best car most often. According to the researchers, this is because people can digest much more information with the subconscious part of their brain than with the conscious. We can think consciously about the acquisition of gloves or a bicycle bell. But when it gets more complicated we are better off leaving it to our subconscious. Sleeping on it for a night gives the subconscious the time to calculate and the next day you will know what's best.

Bos, M. W. & Dijksterhuis, A. 'On making the right choice: the deliberation-without-attention effect' in *Science* 311, 9 (2006), pp. 1005–1007

77.

MEN ONLY
DON'T LOOK AT BEAUTIFUL WOMEN (IF YOU STILL HAVE THINKING TO DO)

Travelling by train and sitting next to a lady with beautiful long legs? You'd better postpone your important decisions.

Psychologists at McMaster University in Hamilton, Canada, found out that men change into short-sighted creatures after seeing beautiful women. In an experiment, the researchers offered 96 male and 113 female volunteers a financial choice. Would they want to receive a certain amount of money today, or would they prefer to get a higher sum at a later date? It was already an accepted fact that people will be satisfied with less if they don't have to wait: many of us would rather receive €99 today than €100 next month. Still, in other situations it can be useful to wait: €25 today or €100 next month? Most people will wait. Some people, though, are extremely sensitive to a quick result. Heroin addicts, for example, are not very patient. They need money for their next shot, and will therefore turn down higher amounts of money in the future. And what was the conclusion of the Canadian research? The average man resembles a heroin addict. At least, when he is shown a couple of pictures of beautiful women right before taking financial decisions.

The male instinct probably causes the trouble. The pictures of beautiful women activate the parts of the male brain that deal with reproduction. In order to be able to conquer a woman, men could use some extra pocket money. That is why they will opt for short-term gain. And did women suffer from the same problem when shown pictures of beautiful men? No, not at all.

Daly, M. & Wilson, M. 'Do pretty women inspire men to discount the future?' in *Proceedings of the Royal Society B: Biological Sciences* 271, *Biology Letters Supplement* 4 (2004), pp. S177–S179

78.

DON'T GO TO THE LARGEST SUPERMARKET

The prospect of an enormous supermarket in which you could lose your way might seem exciting, but does such an overwhelming choice make us happy?

To look into this researchers from Columbia University and Stanford University in the USA conducted an experiment.

They found 134 chocolate lovers ready to come and taste chocolate. One section of the volunteers was given chocolates in six different flavours. Others were given more choice: they were presented with at least 30 flavours. Then the researcher told a little lie. 'We're doing a marketing research study that examines how people select chocolates. What I would like you to do is take a look at the names of the chocolates and the chocolates themselves, and tell me which one you would buy for yourself.' The volunteers chose one flavour and then had to judge it. The tasters who could chose from 30 different flavours had more fun during the experiment but were in the end less satisfied with their choice than those with only six choices. Why was this? Simple. The more choice there was, the greater the doubt as to whether there may have been a much tastier chocolate.

Iyengar, S. S. & Lepper, M. R. 'When choice is de-motivating: can one desire too much of a good thing?' in *Journal of Personality and Social Psychology* 79, 6 (2000), pp. 995–1006

DON'T WIN THE LOTTERY

Luck is a relative concept. The question of just how relative was investigated by Philip Brickman in a splendid experiment in 1978. Using a questionnaire the American social psychologist from Michigan University analysed the personal feelings of luck in a group of very lucky people – 22 lottery winners. For a comparison he used a group of 29 extremely unlucky people, all of whom had been maimed in accidents. Both groups had 1 to 12 months to get used to their situation before the research began. Brickman also set up a control group, made up of people who had a 'normal' amount of luck in their lives.

The results were surprising. The lottery winners were a little happier in their lives than the maimed people, but they scored significantly worse than the control group, those of average good luck. These results are probably down to the contrast effect. When someone wins a large sum of money, the experience is so extremely positive that other nice things in daily life lose their worth.

Brickman, P., Coates, D. & Janoff-Bulman, R. 'Lottery winners and accident victims: is happiness relative?' in *Journal of Personality and Social Psychology* 36 (1978), pp. 917–972

aSK FOR A RECTAL MASSAGE (AND GET RID OF THOSE HICCUPS)

THIRTEEN TIPS TO OVERCOME PAIN AND OTHER DISCOMFORTS

80.

DON'T BLOW YOUR NOSE

Runny nose? Try not to blow it! According to research from the University of Virginia this could only complicate matters. Researchers inserted fake snot in the nasopharynx of four healthy volunteers. Then they were asked to blow their noses at full force, while lying on their backs. After that, the scientists used a CT-scan to follow the travels of the nasal fluid. They discovered that, due to the sudden increase in air pressure when nose blowing, the snot located in the mucus doesn't end up in the handkerchief, but in the sinuses.

If this had been real snot, it would have been full of viruses and bacteria that could easily worsen a cold. The researchers repeated the experiment with sneezing and coughing, but the air pressure produced during these actions was much less. Almost no snot at all was found in the sinuses. So what to do when you get a runny nose? Sniffing the mucus back into your mouth might be a better solution. And if you can't stop yourself from blowing your nose: make sure to do it gently, without making too much noise.

Gwaltney Jr, J. M., Hendley, J. O. & Phillips, C. D. 'Nose blowing propels nasal fluid into the paranasal sinuses' in *Clinical Infectious Diseases* 30 (2000), pp. 387–391

81.

THINK: THIS WON'T HURT

Just about to get a tattoo? Or hop into a steaming hot bath? Then say out loud: 'This is not going to hurt at all.' The amount of pain we feel is highly dependent on what we are expecting, as shown by American research from Wake Forest University School of Medicine. Ten brave volunteers were blindfolded and had a hot rod repeatedly pressed against their calves. It heated a small area of flesh to a temperature of 46, 48 or 50 °C; hot enough to cause serious pain, but not hot enough to burn them. The experiment began with practice sessions. During these sessions the pauses between pain inflictions were established as: 7.5 seconds for 46 °C, 15 seconds for 48 °C and 30 seconds for the 50 °C. So the volunteers learned what they could expect: the longer they had to wait the heftier the pain. The intensity of the pain felt was assessed through a questionnaire. Brain scans also measured the activity of the pain centre in the brain. Two days after this session the volunteers again received a series of pain inflictions. But this time the waiting period didn't always match the temperature. If the highest temperature was 'secretly' used after 7.5 seconds this caused much less pain than in the first session, purely because a smaller pain infliction was expected. The reduction in pain was substantial, comparable to the effect of a small dose of morphine.

Coghill, R. C., Koyana, T., Larienti, P. J. & McHaffie, J. G. 'The subjective experience of pain: where expectations become reality' in *Proceedings of the National Academy of Sciences of the United States of America* 102, 36 (2005), pp. 12950–12955

PLAY THE DIDGERIDOO (AND SLEEP BETTER)

Alex Suarez from Switzerland was often tired during the day. At night he snored like a trooper, so loudly that his wife was often kept awake by it. Then one night she was waiting for the next snore but to her shock it didn't come: Suarez was no longer breathing. She roughly shook him awake and saw that he was still alive. In a sleep laboratory in Zürich sleep apnoea was diagnosed, a condition which causes the sufferer to stop breathing every now and then during sleep. The body is deprived of oxygen so the sleeper wakes with a start. With Suarez that happened no less than 17 times an hour. He was given a sleeping mask, attached to a respirator. But because the mask was uncomfortable, he began an experiment: he started playing the didgeridoo, an Aboriginal Australian instrument. To get music out of one of these painted hollow tree trunks a circular breathing pattern is needed: you have to breathe in through your nose and simultaneously out through your mouth. The cheeks work as a pair of bellows.

Suarez hoped that he could stop his snoring by training his breathing. It turned out to be a good move. After half a year Suarez had no more problems. Did it have something to do with his new instrument? Sleep experts could barely believe it. Still, they recruited several snorers. They were given plastic didgeridoos and four lessons from Suarez. Then they had to play at least five days a week for four months. And it helped! After this intensive breathing training the didgeridoo players were less sleepy during the day than before. And what's more, their partners complained less about them snoring.

Lo Cascio, C., Puha, M. A. & Suarez, A. 'Didgeridoo playing as alternative treatment for obstructive sleep apnoea syndrome; randomised controlled trial' in *British Medical Journal* 332 (2005), pp. 266–270

SUFFERING FROM DIARRHOEA? EAT CHOCOLATE

Sixteenth-century explorers knew that a couple of cacao beans can work wonders when it comes to forming firm stools. Now, at long last, science understands why this works. One of the main culprits that causes diarrhoea is a protein known as CFTR that regulates fluid secretion in the small intestine. With too much CFTR you suffer from 'the runs'. Is there a way to stop this from happening? Yes, says the Children's Hospital Oakland in the USA. In their lab researchers noted how an important constituent of chocolate, flavonoids, attached themselves to CFTR and so stopped the runs. The darker the chocolate, the greater the effect.

Schuier, M. & Sies, H. 'Cocoa-related flavonoids inhibit CFTR-mediated chloride transport across T84 human colon epithelia' in *Journal of Nutrition* 135 (2005), pp. 2320–2325

KISS! (AND HELP FIGHT HAY FEVER)

Red eyes, runny nose, sneezing fits – hay fever sufferers get these symptoms every year. But luckily Doctor Hajime Kimata of the Japanese Satou Hospital has found a pleasant way to ease the suffering: kiss!

He asked 24 hay fever patients to kiss their partners for half an hour, to the tune of love songs like 'When you wish upon a star' and 'Can you feel the love tonight?'. Just before kissing and just afterwards the patients gave a blood sample. The results? The amount of histamine in the blood after kissing was markedly lower. And that just happens to be the hormone that causes hay fever. It is possible that this improvement in health is thanks to the relaxing effect of kissing. Other researchers have shown that stress increases the amount of histamine the body makes. But this theory is not entirely convincing. Cuddling, which after all also relaxes, did not reduce histamine levels, a second experiment conducted by Kimata showed. So for the moment, hay fever sufferers will just have to kiss a lot.

Kimata, H. 'Kissing selectively decreases allergen-specific IgE production in atopic patients' in *Journal of Psychosomatic Research* 60 (2005), pp. 545–547

85.

GOT A HEADACHE? TRY ACUPUNCTURE

Pushing needles into your flesh is the solution to all your problems. It seems hard to believe but recent research does suggest that acupuncture really can help. Just don't ask why!

Researchers from the Technical University in Munich were trying to discover why when they performed a study of patients in 28 German hospitals, in which 270 patients with chronic headaches were split into three groups:

- Group 1 – were put on a waiting list.
- Group 2 – were given, over two months, 12 acupuncture sessions using the traditional Eastern method.
- Group 3 – received the same treatment but with the needles placed very superficially and in any old spot.

All the patients kept a diary of how often they suffered from their headaches. The researchers compared their diaries for four weeks leading up to the treatment and then four weeks after the treatment. What did they find? For the patients who received 'real' acupuncture, the number of days they had headaches was reduced by seven. The patients on the waiting list were less lucky – they noted just one and a half fewer headache days. But here's the strange thing: the patients who had the fake acupuncture also noted seven fewer days with headaches, the same as those who had the real acupuncture. Comparable research with migraine sufferers gave the same results: it would appear that it doesn't matter where or how deep you put the needles in the skin.

The healing power of acupuncture could be explained by the placebo effect. This claims that someone can be cured with fake pills and hoax treatments, as long as the patient believes they work. So if you go to the acupuncturist for help with getting rid of a headache, you'd better have faith in the power of needles.

Linde, K. 'Acupuncture for patients with migraine: a randomised controlled trial' in *Journal of the American Medical Association* 293, 17 (2005), pp. 2118–2125

86.

WHEN FEELING FAINT: CROSS YOUR LEGS, FLEX YOUR MUSCLES

Once in a blue moon, a miracle cure comes along out of nowhere. A 33-year old patient with a nervous system disorder explained what he always did if he felt he was about to faint: he crossed his legs and flexed his muscles. If he kept this up long enough, at least half a minute, the faint feeling would often go away.

Researcher Nynke van Dijk decided to see whether this trick worked for others too. She invited 21 notorious fainters to her lab in the Academic Medical Center in Amsterdam, the Netherlands. There they had to lie on a tilting table, one that can be rotated so that the patient is put in a vertical position, head up. In the meantime she turned the heating up and gave the patients a pill to put under their tongues, 'hoping' that they would thereby pass out. Once the fainters began to feel themselves getting light-headed, Van Dijk urged them to cross their legs and flex their muscles, and to keep doing it.

It appeared to work: for all the patients, the feeling of being about to faint went away.

So how does this miracle cure work? Simple. Fainting is a reflex, during which the blood vessels in the legs and stomach open right up. Thereby the blood sinks downwards, creating a shortage of oxygen in the brain and causing the patient to pass out. By flexing the muscles before this moment, the blood that has sunk into the blood vessels is pushed up into circulation again. The positive consequence: the heart is helped with pumping the blood round the body and the brain receives oxygen again.

van Dijk, N. 'Studies on Syncope' (2006), PhD thesis at the University of Amsterdam

87.

DON'T USE DUCT TAPE ON WARTS

There was big news in 2002: warts could apparently be dealt with using a product every do-it-yourselver has in the shed – duct tape, a wide and self-adhesive sticky tape that is usually silver in colour. This conclusion was reached by researchers of the children's hospital in Cincinnati America, but does it really hold? Researchers at the University of Maastricht, the Netherlands, thought not. They had found a few flaws in the research and decided to look into the whole thing again.

At three primary schools they found 103 children with warts. Half of them were instructed to put duct tape over the warts for six weeks. They could use fresh tape every week. The other children were given a placebo: they had to put a ring shaped plaster around their wart one night a week. After six weeks the researchers counted the number of warts on the children. They also measured the size of the wart in question and a maximum of four surrounding warts. The conclusion: the duct tape doesn't help. And what's more, nearly all the children said that the tape wouldn't stay on and seven complained of itchiness and rashes.

Haen, M. & de Spigt, M. G. 'Efficacy of duct tape vs placebo in the treatment of verruca vulgaris (warts) in primary school children' in *Archives of Pediatrics and Adolescent Medicine* 160 (2006), pp. 1121–1125

88.

WATCH A FUNNY FILM (AND EASE THE PAIN)

Is that headache bothering you again? Try watching a comedy – it might ease the pain. Researchers from Heinrich-Heine University in Germany asked 56 women to put their hand in a bucket of water that was kept at body temperature. Afterwards they had to submerge their hand in a bucket of ice cold water, for as long as they could, until the pain became unbearable.

The same experiment was repeated, but now after watching an episode of *Mr Bean*. (To be precise: the episode where the comedian brushes his teeth while driving a car, and sticks his head out the window to wash his mouth with a windscreen washer.) Even twenty minutes after watching the movie, the participants could keep their hand submerged in the ice water much longer than in the first experiment. The researchers concluded that laughing can work as a pain killer.

Ruch, W., Velker, B. & Zweyer, K. 'Do cheerfulness, exhilaration and humor production moderate pain tolerance?' in *International Journal of Humor Research* 17 (2004), pp. 67–84

ASK FOR A RECTAL MASSAGE AND CURE THOSE HICCUPS

In 1988, in *Annals of Emergency Medicine*, a remarkable letter was published. Francis Fesmire of the academic hospital in Jacksonville, Florida, wrote about meeting a 27-year-old man who had had the hiccups for 72 hours. When he arrived at the hospital he was hiccupping at an impressive rate of 30 times a minute. In all other aspects the man seemed healthy.

The doctor pulled the patient's tongue, and pushed down on his eyeballs, all to stimulate the out-of-control vagus nerve ('the Wanderer' or 'the Rambler') in order to stop the hiccups. Nothing helped. Then Fesmire remembered that he had once read about 'digital rectal massage' (that's digital as in relating to the finger, not to computers!). The treatment basically involves massaging the anus in a slow circular movement with a finger and supposedly helps reduce heart palpitations. Because that problem is comparable with persistent hiccups, Fesmire decided to go for it. And it worked! To the patient's great relief the frequency of the hiccups was immediately reduced and within 30 seconds he was completely free of them. Coincidence? No way. Two years later Israeli researchers published a similar case: their 60-year-old patient was cured of hiccups after a relaxing massage.

Fesmire later found another solution for people with intractable hiccups: good sex. An orgasm also stimulates the nervus vagus, the doctor claims.

Fesmire, F. M. 'Termination of intractable hiccups with digital rectal massage' in *Annals of Emergency Medicine* 17, 8 (1988), p. 872

Bassan, H., Odeh, M. & Oliven, A. 'Termination of intractable hiccups with digital rectal massage' in *Journal of Internal Medicine* 227, 2 (1990), pp. 145–146

BE OPERATED ON IN THE MORNING

Of course, mostly you don't get the choice. But perhaps one day a friendly doctor will appear by your bedside and ask you what time of day you would prefer to be operated on. Don't hesitate – ask for the morning slot. Research by Melanie Wright of Duke University Medical Center in Durham, North

Carolina, shows that during that part of the day less mistakes are made.

She analysed the data of 90,159 operations performed between 1 May 2000 and 4 August 2004 in her own hospital. The least sprightly were those who had lain on the operating table between 3 and 4 in the afternoon. They complained most often of nausea, vomiting and pain after the procedure. This is probably the surgeon's fault, Wright suspects. At the end of their shift they are tired and therefore do less careful work. To minimise the risk patients should book themselves in for surgery at 9 a.m.

Phillips-Bute, B. & Wright, M. C. 'Time of day effects on the incidence of anaesthetic averse events' in *Quality and Safety in Health Care* 15 (2006), pp. 258–263

ASK FOR A ROOM WITH A VIEW (AND RECOVER MORE QUICKLY)

The two double rooms were identical. Both were on the same floor in a small hospital in Pennsylvania. The patients lying in them were even looked after by the same nurses. But there was one difference: one room looked out on a copse of trees, the other on a concrete wall. Researcher Roger Ulrich of the University of Delaware, USA, wondered whether this made a difference. He decided to analyse the data of patients who had occupied the rooms between 1972 and 1981. To make the comparison fair he only chose people who had exactly the same gallbladder operation. The results were remarkable. Those looking at trees spent an average of eight days in the hospital, whereas those looking at the wall on average stayed a day longer. Ulrich also took note of the nurses comments like 'needs a lot of encouragement' or 'is over-wrought and cries'. For those with a nice view these comments were far fewer: on average one per person compared to four for those who had a wall to look at. Were there any other differences? Definitely. In the rooms with a view fewer heavy painkillers were swallowed by the patients.

Ulrich, R. 'View through window may influence recovery from surgery' in *Science* 224, 4647 (1983), pp. 420–421

WHEN STRESSED, HOLD A MAN'S HAND

How do you stimulate stress in a lab? Easy – by fixing an electrode to someone's ankle and giving them a little electric shock every now and then. Sixteen married woman volunteered for an experiment conducted at the University of Virginia. One by one they had the electrode bound around their ankles and took their position in a brain scanner. A screen was placed in front of their faces. If a red 'X' appeared on the screen, the woman had a 20 per cent chance of receiving a shock. If a blue 'O' appeared then she would definitely not get a shock. Not entirely unexpectedly, the brain scans showed that the 'X' caused more stress than the 'O'. This was shown by more brain activity in areas of the brain such as the hypothalamus, which plays an important role in the processing of emotions. However, the researchers found that the level of stress could be reduced if the woman held the hand of a man, even he was a total stranger to her. But the calming effect only became really marked if she held her husband's hand.

The question remains: are men also calmed by holding hands? Alas, the researchers didn't look into that.

Coan, J. A., Davidson, R. J. & Schaefer, H. S. 'Lending a hand: social regulation of the neural response to threat' in *Psychological Science* 17, 12 (2006), pp. 1032–1039

WASH YOUR HANDS (CLEANSE YOUR SOUL)

FIVE TIPS TO GET RID OF YOUR TRAUMAS AND OTHER PSYCHOLOGICAL PROBLEMS

WRITE YOUR WAY OUT OF IT

Bad news for the psychologist's wallet: you can climb out of an emotional dip by simply writing about it.

James Pennebaker, professor of psychology at the University of Texas, has over the years asked hundreds of people to try writing therapy. That therapy consisted of 15 to 30 minutes of writing a day over the course of three to five consecutive days. Participants had to write about their deepest feelings and most painful memories. There was no commentary on the texts; they were just writing for themselves.

For every experiment Pennebaker also asked a control group to write about neutral subjects, for example their plans for the day. This way the professor showed how well the emotional writing therapy worked – participants felt better afterwards. Unemployed people who put their soul's torments down on paper got a job more quickly than those without work who wrote about neutral subjects. Lovers who had written their troubles away stayed together longer than those who wrote about everyday things. Students who poured their emotions out on paper got higher grades than students who gave a business-like run down of their daily activities.

Pennebaker, J. W. 'Writing about emotional experiences as a therapeutic process' in *Psychological Science* 8 (1997), pp. 162–166

NEVER BELIEVE YOUR HOROSCOPE

What plays a defining role in the formation of our characters? Is it our genes? Our upbringing? Or is the position of the sun, moon and other heavenly bodies? Astrologers claim the latter; birth horoscopes are based on the map of the stars on the day of our birth.

Peter Hartmann of Aarhus University, Denmark, wanted to finally establish if there was any truth in all this star witchery. The Danish researcher compared birth date to the results of personality and intelligence tests for 15,000 volunteers. The data was mostly taken from the National Longitude Study of the Youth, an American database for which volunteers filled in questionnaires for several years. From this data you can, for example, tell whether someone is particularly introverted or extroverted.

The results were somewhat sobering. Our birth date (and thus the positioning of the stars and the planets at that moment) tells us absolutely nothing about our intelligence or our character. So why do people so often recognise themselves in the wise words of astrologists? The reason for that is simple. We readily recognise ourselves in general descriptions. Take this sentence: 'You enjoy being amongst people, even though you can sometimes feel very alone.' It seems a profound and utterly personal observation but it is applicable to pretty much the whole world population.

Hartmann, P. 'The relationship between date of birth and individual differences in personality and general intelligence: a large scale study' in *Personality and Individual Differences* 40 (2006), pp. 1349–1362

WASH YOUR HANDS (CLEANSE YOUR SOUL)

'Washing away your sins' is a ritual known in many religions. But does cleaning the body really have a healing effect on the mind?

The University of Toronto asked volunteers to either write down a positive deed from their past (like helping a co-worker) or write about a negative deed (like sabotaging a co-worker). The participants were told that they were taking part in a study on handwriting and character, so they were unaware of the real reason for the experiment. After the writing sessions the volunteers could choose between two gifts: a wet tissue to clean their hands, or a pencil. The results were intriguing. Of the people who had just showed a bad side of themselves, 67 per cent picked the tissue. Of the people who had written down a positive deed, only 33 per cent went for the tissue. These results suggested that committing what is perceived to be a sin leads to the urge to wash oneself.

But does this cleansing really have the desired effect? The researchers started a new experiment, with new volunteers. This time the test subjects only wrote about negative things they had done in the past. Afterwards they had the opportunity to wash their hands, but only if they wanted to. Finally, the volunteers were asked if they would like to participate in another study. It would be unpaid but, so they were told, they would really help out a desperate graduate student.

Many more of the volunteers with 'dirty hands' offered to help than the ones with 'clean hands'. This was exactly what the researchers had expected: participants who had cleaned their hands were less motivated to volunteer because the sanitation wipes had already washed away their moral stains. The participants with dirty hands needed other ways to restore a suitable moral self and decided to volunteer again.

Liljenquist, K. A. & Zhong, C. 'Washing away your sins: threatened morality and physical cleansing' in *Science* 313 (2006), pp. 1451–1452

STOP PRAYING FOR OTHERS (IT DOESN'T HELP ANYWAY)

Is there any point in praying for a sick neighbour? Definitely, said the deeply religious Christians of the John Templeton Foundation. The American organisation decided to conduct serious research to show that science could prove that praying has a positive effect on the curing of illnesses. The cost: 2.4 million US dollars. No less than 1,802 heart patients who had undergone a by-pass operation between 1998 and 2000 were divided into three groups:

- Group 1 – These people were prayed for, without them knowing it.
- Group 2 – This group was not prayed for. They weren't aware of this either.
- Group 3 – The patients in this group did receive prayers and they knew about it.

By comparing the first two groups the researchers could see if prayer itself helped. The third group was added to see if there were perhaps any psychic influences; maybe just the thought of being prayed for would work positively.

The praying took place in three churches: in Minnesota, Massachusetts and Missouri, all far from the hospitals in which the patients were treated. The people praying were given the first name and the first letter of the last name of the patient and were asked to pray for fourteen days for a 'successful surgery with a quick, healthy recovery and no complications'. Did the prayers work? No. Between the first two groups, those who didn't know whether they were being prayed for or not, there was no difference. In both groups roughly half suffered from complications. The surprise was in Group 3: those who knew they were being prayed for suffered more complications that the other patients. Why was this? The researchers suspect it is coincidence.

Benson, H., Dusek, J. A. & Sherwood, J. B. 'Study of the therapeutic effects of intercessory prayer (STEP) in cardiac bypass patients: a multicenter randomised trial of uncertainty and certainty of receiving intercessory prayer' in *American Heart Journal* 141, 4 (2006), pp. 934–942

97.

INSECURE? READ BAD DETECTIVE NOVELS

'Murder because of lust or greed?' so stated the title of a thriller given to 84 students to read. They read about a businessman who is found dead by his cleaner in his villa. It is obvious that he is the victim of a brutal murder. Two women have a motive, both lack an alibi: the businessman's wife and his mistress. Who did it? To establish what kind of stories people enjoyed, researcher Silvia Knobloch-Westerwick of Ohio State University presented the students with different versions of this story:

- *Climax* – some readers were initially given an equal number of clues pointing to the wife and the mistress. His wife turned out to be the culprit.

- *Confirmation* – other readers were particularly given clues that the mistress did it. And she had in fact done it.
- *Surprise* – the last group were given at the beginning of the story the impression that the mistress had done it. In this version an unexpected ending followed: the business man's wife turned out to be guilty after all.

The first version was given the most worth, as all was open until the end of the story. What was surprising was how insecure people got the most enjoyment out of stories that confirmed their suspicions. That gave them a good feeling. For people with a lot of self-confidence the opposite was true: they would rather be surprised.

Keplinger, C. & Knobloch-Westerwick, S. 'Mystery appeal: effects of uncertainty and resolution of the enjoyment of mystery' in *Media Psychology* 9 (2006), pp. 193–212

VICIOUS DOG? DON'T TRUST THE OWNER

NINE TIPS THAT ARE CERTAINLY USEFUL, BUT DID NOT FIT INTO THE OTHER CHAPTERS

SIT IN THE MIDDLE
(AND SEEM MORE INTELLIGENT)

In the TV series *The Weakest Link*, eight perspiring candidates stand in a row. Who will be sent off after a round of questions? Quite often it is the candidate who has performed the worst. After all, this person has earned the least amount for the group. But Priya Raghubir of the University of California suspected that elimination didn't just depend on performance.

To test her suspicions, she watched 20 episodes of the quiz show. She paid attention to the peripheral contestants (in positions one and eight in the row) and on the two players in the centre (positions four and five). What did she find? In 45 per cent of the programmes one of the centre players won the final. The players at the ends of the row rarely went home with the prize money: just in one in ten episodes. According to the researchers that is because we expect important people to stand in the middle of a group. Take the winner of a gold medal, who is surrounded by the runner up and the bronze medal winner. Or look at a singer, standing in the middle of his band on the podium. Because of this we are inclined to pay less attention to how they actually perform.

Raghubir, P. & Valenzuela, A. 'Centre-of-inattention: position biases in decision-making' in *Organisational Behaviour and Human Decision Processes* 99 (2006), pp. 66–80

VICIOUS DOG? DON'T TRUST THE OWNER

So your neighbour has a pit bull terrier? Maybe he's not the right person to ask to look after your plants while you're on holiday.

Researchers from different institutions in the American state of Ohio found out that owners of vicious dogs like pit bulls, Rottweilers and Akitas are not so nice themselves. The researchers studied the criminal convictions of 355 dog owners. They looked at big crimes like domestic violence or fights with fire arms, but also at smaller crimes like traffic offences and driving under the influence. The conclusion was amazing. Every owner of an aggressive dog had at least one criminal conviction. For people with normal dogs this was only six in ten. Crimes involving drugs were committed eight times more often by owners of vicious dogs, crimes involving arms seven times more often and domestic violence 2.5 times more often.

Barnes, J. E. & Boat, B. W. 'Ownership of high-risk ('vicious') dogs as a marker for deviant behaviors' in *Journal of Interpersonal Violence* 21, 12 (2006), pp. 1616–1634

DON'T SPEAK DIALECTS WITH YOUR CHILDREN

People who speak a dialect don't get as far as people who speak the generally accepted form of the language. This is shown in research by Gerbert Kraaykamp of Radboud University in Nijmegen, the Netherlands.

The researcher asked a representative group of more than three thousand Dutch of all ages if they had spoken dialect with their parents and friends when they were younger. At least 43 per cent of the people said they had. Then Kraaykamp asked the people what their highest level of education was and about their first and last job. Finally, he assessed their vocabulary.

The result? Those people who had spoken a dialect in their youth appeared to have enjoyed an average of half a year's less education, their jobs were lower down the social ladder, and what's more, they scored worse on the vocabulary test. This is partly explainable by the social background of the dialect speakers: those with less education often speak dialects, and the children of the less educated have less chance of higher education. But this wasn't the only explanation. Even when Kraaykamp corrected the results for lower social class, dialect speakers scored on average worse.

Kraaykamp, G. 'Dialect en sociale ongelijkheid: een empirische studie naar de sociaal-economische gevolgen van het spreken van dialect in de jeugd' in *Pedagogische Studiën* 82 (2005), pp. 390–403

101.

NEVER TRUST ANYONE WITH YOUR SECRETS

Imagine you tell a good friend that you have committed adultery. Will he keep it a secret? Probably not, research by Veronique Christophe of Lille University tells us. The social psychologist presented 134 students aged 18 to 28 years old with a list of questions. First the volunteers noted whether they had recently been entrusted with an important emotional event by an acquaintance. Then they had to say whether they had spoken to anyone else about it. Sixty-six per cent admitted having done this, 39 per cent the very same day.

Why are we so bad at keeping other people's secrets? A possible explanation is social visibility: whoever tells a secret can count on a lot of attention from the person to whom he is speaking. But let's not judge the loose tongues and gossipers too harshly, because receiving shocking personal information can really unsettle you. This gives rise to a natural need to offload on someone else.

Christophe, V. & Rimé, B. 'Exposure to the social sharing of emotion: Emotional impact, listener responses and secondary social sharing' in *European Journal of Social Psychology* 27 (1997), pp. 37–54

102.

GO LAST (AND HAVE MORE CHANCE OF WINNING)

Taking part in *The X Factor* or the *Eurovision Song Contest*? Or in a jury sport like gymnastics or figure skating? Make sure you perform last.

The results of various research projects suggest that those who go last are unconsciously favoured by the judges. Wändi

Bruine de Bruin of Carnegie Mellon University, Pennsylvania, studied the results of international figure skating competitions between 1994 and 2004. In the first round the order of the competitors was decided by lottery. Those who started later on average scored higher.

The explanation for this phenomenon is unknown. Perhaps the judges are inclined to concentrate on the positive aspects which dissociate a performer from their predecessor, resulting in rising scores.

Bruine de Bruin, W. 'Save the last dance for me: unwanted serial position effects on jury evaluations' in *Acta Psychologica* 118 (2005), pp. 245–260

Bruine de Bruin, W. 'Save the last dance II: unwanted serial position effects in figure skating judgements' in *Acta Psychologica* 123, 9 (2006), pp. 299–311

103.

MAKE A LOT OF FRIENDS (AND LIVE LONGER)

Hoping to grow old? Then start building up an impressive circle of friends. Australian research shows that people with a large social network die later than those who have gathered fewer people around them. Not family but friends and acquaintances seem to be essential. Lynne Giles and colleagues from Flinders University in Adelaide followed 1,500 Australian septuagenarians for ten years. At the end of the study some 900 had died. The survivors had on average more friends than the deceased. It appeared that the more friends the subjects of the study had, the greater their chance of living long. Why this is, exactly, is unclear. Maybe having friends makes people smoke and drink less, and their friends make them go to the doctor if they are ill, the researchers suggest. Friendship could also prevent depressions and is good for self-confidence.

Giles, L. C. & Glonek, G. F. V. 'Effect of social networks on 10 year survival in very old Australians: the Australian longitudinal study of aging' in *Journal of Epidemiology and Community Health* 59, 7 (2006), pp. 574–579

DON'T GOSSIP

Look out! What you say about others can come back at you like a boomerang. John Skowronski of The Ohio State University in the USA asked 47 students to look at mugshots of strangers. Under each photo was printed a line the person had said about another. For example, under the face of a young woman: 'He hates animals. Today he was walking to the store and he saw this puppy. So he kicked it out of his way.' Two days later the volunteers were again shown the mugshots, this time without the accompanying text, and asked to comment on the person's character. Surprisingly, the woman who gossiped about the animal hater was now classed as 'cruel' herself. Apparently, we project the content of the message on the messenger. This effect was repeated when the experiment was performed with video messages, and also when the volunteers were repeatedly told that the utterances had nothing to do with the photographed people. Skowronski thinks that dirty ad campaigns, where competitors are smeared, are counter-effective.

But hold on a minute. What happens when you say something nice about someone else? For example, that the boy in the corner is a great tennis player? Unfortunately, even though your words will stick to you, it won't actually make you a great tennis player.

Carlston, D. E., Mae, L. & Skowronski, J. J. 'Spontaneous trait transference: communicators take on the qualities they describe in others' in *Journal of Personality and Social Psychology* 74, 4 (1998), pp. 837–848

AVOID CHURCH OVER CHRISTMAS

Stephan Weber of Duisburg-Essen University decided to use the last days of 2004 to answer an important question: is the air in the Catholic Church detrimental to health? In the Sint Engelbert Church in Mulheim an der Ruhr he installed apparatus to count the tiny, invisible smoke particles that float in the air. Weber left the sensors measuring for two weeks. What did he find? The amount of dust in the church was reasonably low during normal services. But on Christmas Eve, Christmas Day and New Year's Eve, Weber discovered that there was seven to nine times more dust in the air. During these gatherings, not only were all the candles lit but there was also incense burning. The latter turned out to be a big polluter. The number of micro particles in the church was even higher than the air next to busy motorways. Dangerous? Definitely. Inhaled in large quantities dust can lead to respiratory problems, and lung, heart and vascular diseases.

Weber, S. 'Exposure of churchgoers to Airborne Particles' in *Environmental Science & Technology* 40, 17 (2006), pp. 5251–5256

STOP READING (AND START ASKING QUESTIONS)

Just one hour left before that tricky theory exam for the diving school, driving licence or evening class. What to do: read the textbook or go through some practice questions? The latter is more sensible, research from Washington University in St Louis shows. Students were given 25 minutes to read a boring biological text about the toucan, a bird found in Mid and South America. They then handed the text back and were split into three groups.

- Group 1 – was allowed to go straight home.
- Group 2 – had to answer questions about the material like 'Where do toucans sleep?' They were given 25 seconds per question and weren't told what the correct answer was.
- Group 3 – was given more information about the toucan to read. Without them knowing it, the text they were given to read was point by point the answers to the questions given to Group 2. So one of the points was: 'Toucans sleep in holes in trees.'

The following day all the students were tested. They were given completely new questions. And who scored best? Group 2 answered nearly 10 per cent more questions correctly than the other two groups. Conclusion: always use part of your study time to answer questions. Even if the questions don't reappear in the exam, you will score higher marks than if you just read.

Chan, J. C. K. & McDermott, K. B. 'Retrieval-induced facilitation: initially non-tested material can benefit from prior testing of related material' in *Journal of Experimental Psychology* 135 (2006), pp. 553–571